MDM
模 具 設 計 與 製 造
Mold Design & Manufacturing

電腦輔助工業
產品設計及實作

Computer Aided Industrial
Product Design and Practice

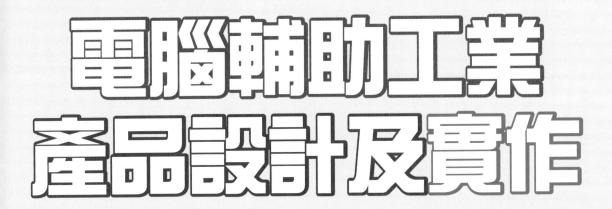

王松浩
尤瑞雅
貝德杰 著

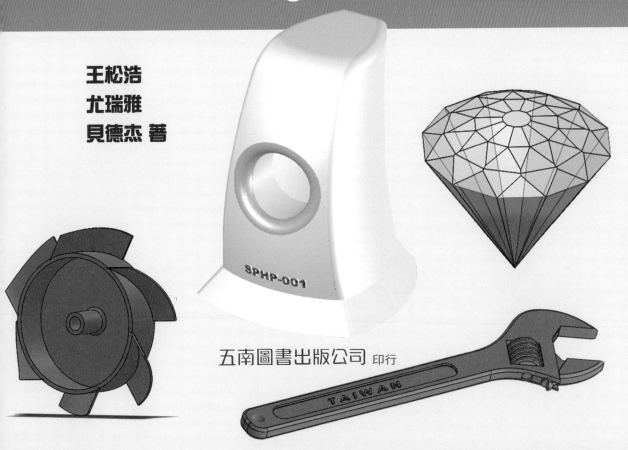

SPHP-001

五南圖書出版公司 印行

導言

　　工業產品的設計中，工學、人體工學和美學為三大基本要素。本書將介紹基本的工業產品設計概念與理論。著重訓練學生在人體工學和美學方面的基本知識和技能。運用現代先進的 3D 立體及曲面 CAD 造型，以及配合逆向工程快速原型 RP 技術，使工程類學生日後能夠更加得心應手地融入日益競爭的職場之中，並為臺灣的工業產品設計和品牌發展計畫貢獻力量。本書之重頭戲在於融入了不少比較複雜的實作範例並加以詳盡敘述，試圖使讀者嘗試無師自通的感覺且頗有成就感，也會使已經具備一些 3D 繪圖基礎技能的讀者感到如虎添翼。此外本書還盡量以中英文雙語進行論述，以適應全球化和外籍生增加的需求。

Introduction

　　For Industrial Product Design (IPD), mechanics, ergonomics and aesthetics are three major aspects. This book will introduce the basic concepts and theories of IPD especially for Engineering Students. Emphasize will be on the training of basic knowledge and techniques of ergonomics and aesthetics. Combining advanced 3D solid/surface modeling on Reverse Engineering and Rapid Prototyping Technology, the book will help engineering students better prepared to face today's more and more competitive market and contribute more on Taiwan's Product Design and Brand Development. Moreover, the book is written in both Chinese and English, to meet the requirement of globalization and foreign students.

有人說，「讓我們提供客戶現在所需要的吧」，但這不是我們的做法。我們的工作是要研發出人們將會想要的產品。

——史蒂夫・賈伯斯

"Some people say, 'Give the customers what they want.' But that's not my approach. Our job is to figure out what they're going to want before they do."

—— Steve Jobs

提供人們所需要的，你可以得到溫飽；
提供人們所想要的，你卻有可能發財。

——猶太商人的座右銘

Sale people what they need, you make a living;
Sale people what they want, you make a milling.

—— A Motto from Jewish Business Men

推薦序 Foreword

OEM（Original Equipment Manufacturer）代工是臺灣企業長久以來的主要經營模式；MIT（Made in Taiwan）也是臺灣企業／產品長久以來的主要標誌。雖然無可非議這些都是臺灣經濟的主要支柱，可是在日益競爭的世界已經遠遠不夠。為迎接知識經濟新時代，臺灣政府在「國家發展重點計畫」下，已將發展品牌列為臺灣產業升級、企業與國家競爭力提升的重要工作。至此臺灣產業轉型關鍵時刻，政府順勢於 95 年度推出「品牌臺灣發展計畫」，並將「全面提升產品形象計畫」納入本計畫內，不僅具體展現政府扶持臺灣企業發展品牌，加速企業國際化與提高市場競爭力的決心；更積極要在國際市場加強宣傳臺灣產業與產品形象，為臺灣企業發展國際品牌營造更有利的生存條件。「品牌臺灣發展計畫」最終目標有二：一是整合資源協助企業建立品牌，營造良好品牌發展環境、二是協助臺灣企業發展國際品牌並提升臺灣國際品牌價值。「品牌臺灣發展計畫」將透過完善品牌發展環境、辦理品牌價值調查、營運品牌創投基金、提升臺灣產品與產業國際形象、建構品牌輔導平臺以及擴大品牌人才供給等措施，扶持臺灣企業發展品牌，達到多元品牌、百花齊放的最終目標。

品牌，就像一個人的信用和名譽，是企業發展不可或缺的資產。品牌對公司而言有兩大好處，第一就是縮短購買產品的時間。當大家都知道你這個公司的時候，買方 Decision Making（做決策）比較快，凡是某某公司出的產品，我就信任你。第二個就是 Conduct Higher Value（創造高價值），像是同樣一個皮包，為什麼香奈兒（Chanel）就賣得比較貴？分析師指出，iPad 基本款每部生產成本只要 290.5 美元，意即蘋果雖然便宜賣，每部可大賺 208.5 美元（約臺幣 6,677 元），毛利超過四成，較高階的 iPad 毛利更超過一半。如果升級為配備 3G 上網功能的 16GB 機型，成本只增加 16 美元，但售價卻貴了 130 美元，毛利率跳升到 52%。與 OEM 業者只有3%、5%的毛利相比，當然就會顯示出品牌價值高太多了。
（聯合報，記者範振光報導，2010年2月3日）

　　要做什麼樣的品牌，是臺灣企業迎戰國際競爭的當務課題。身處後ECFA的新局勢，臺灣國貿局與外貿協會連續五年推動「品牌臺灣發展計畫」精準運用一條龍戰略，帶領臺灣中小企業以優質平價的競爭力，前進新興市場。面對全球經貿環境的瞬息萬變，後ECFA時代的新局勢，臺灣企業在提升應變力、打造競爭力的當下，研發創新和品牌行銷同等重要。要不要做品牌已不是問題，要做什麼樣的品牌和如何建立品牌的牢固地位才是重要的課題。這正是經濟部國貿局和外貿協會竭力推動「品牌臺灣發展計畫」最重要的目標。

　　事實上，臺灣已經具備了很好的條件。首先是臺灣在製造業、服務業的發展，已經累積豐富的經營和全球布局的能力。其次，中國、印度、東協等地新興市場的崛起，讓臺灣的優質平價產品有機會進入，特別是在中國由世界工廠轉入世界市場的好時機。第三則是臺灣業者展現的決心和信心，爭取在世界舞臺發光發熱的機會。

　　這些在國際市場大放異彩的品牌分布各個領域。在資通訊方面，宏碁全球品牌布局成功，宏達電在智慧型手機也不容小覷；華碩切割代工與品牌，進軍全球；鴻海啟動「萬馬奔騰計畫」積極在中國布建通路。在傳統產業方面，捷安特坦言已經很難找到對手，只有不斷自我挑戰；康師傅在中國站穩市場龍頭後，轉進其他新興市場。在通路部分，潤泰集團的大潤發已經成為中國市場數一數二的通路品牌；聯強在3C通路也評價甚高。在飲料市場，85℃咖啡烘培店準備要掛牌上市；除此之外，臺灣最令人讚不絕口的珍珠奶茶，雖然沒有具代表性的品牌，但全世界都有珍珠奶茶的蹤影，極具潛力。

　　美國西北大學「整合行銷傳播之父」Don Schultz博士提到：「品牌行銷不是單行道的促銷，也不是譁眾取寵的廣告口號。建立品牌，必須深入消費者的需求，為最終用戶創造價值，它是連結公司與客戶的橋樑。」對於全球建立品牌的策略，臺灣在建立品牌要走的道路，我們未來如何運用有限的資源來發展全球品牌，都需要躍升的力量和方法。政府期許臺灣成為亞太服務品牌的管理中心，無論是臺資公司打造全球品牌，還是外資企業在臺灣設立區域品牌服務中心，都將使得臺灣發展「多元品牌、百花齊放」的願景指日可待。比如為配合前述發展臺灣國際品牌，經濟部推動 15 年之「全面提升產品形象計畫」納入「品牌臺灣發展計畫」，並將「臺灣精品網站」歸併於「品牌臺灣網站」，整併後面貌清新、使用便捷的新版「品牌臺灣網站」。

　　唯有「Made in Taiwan」這個虛擬品牌還遠遠不夠。而欲求從 OEM 轉型到 OBM（Original Brand Manufacturer），ODM（Original Design Manufacturer）是先決條件，因為必須先有自己設計（Design）的產品才可能建立自己的品牌（Brand）。因此，我們臺灣的技職院校應該在這方面多花一些功夫、多做一些貢獻。讓我們一起努力為臺灣企業培養更多更強的 ODM 人材，成為 DIT（Design in Taiwan）源頭，甚至是 OBM 的源頭，這將是使我們立於不敗之地應該努力的方向。

周煥銘 教授
Huanming Chou, Professor
崑山科技大學副校長兼工學院院長，臺灣
Dean of Engineering School, Kun Shan University, Taiwan

自序 Preface

　　幾乎所有工業產品設計都是 3D 立體的，以往的設計過程中，設計者將頭腦抽象的 3D 立體概念描述到製圖板上 2D 透視，然後又由 2D 的透視翻譯並製成 3D 的實際產品，這個過程不僅十分繁複，更需要百分的專業技巧和訓練。如今隨著電腦軟硬體突飛猛進的發展，產品設計的工具已經不可同日而語。設計者可以非常方便地將頭腦中的 3D 概念，通過 3D CAD 工具直接訴諸於 3D 的虛擬曲面或實體，再由 CNC 或者 RP 工具快速的轉變成可觸、可試甚至可用的實際模型或產品。就如同數位相機的發明和進步使得現今千千萬萬的攝影師應運而生一樣，從一個工程師轉變成產品設計師的門檻也大大降低了。此外由於網際網路發展引起的訊息大爆炸，使得任何昨天的老品牌和老產品；今天的新發明和新產品；以及明天的大趨勢和可能性，靠手指輕輕一點就盡覽於眼下。

　　鑒於以上思考，筆者在五年以前就開始著手準備在崑山科技大學機械系開設「工業產品設計」的課程，鼓勵學生嘗試進入「產品設計」這個「高貴的殿堂」。但是在準備和文獻搜索過程中，雖然各類設計和產品設計書籍琳琅滿目，適合工程類學生在較短時期內完成的學習材料就很難找到了。這是因為全程的產品設計專業科目，大約有色彩與設計（Color & Design）；形狀的發展（Form Development）；設計問題的解決（Design Problem Solving）；工業設計的方方面面（Perspectives for Industrial Design）；工業設計的歷史（History of Industrial Design）；設計繪圖 1（Design Drawing 1）；設計繪圖 1（Design Drawing 2）；設計繪圖 3（Design Drawing 3）；模型製作 1（Model-making 1）；模型製作 2（Model-making 2）；數位影像（Digital Imaging）；3D 成型技術（Introduction to 3D Modeling）等 12 門以上獨立的課程，每門課程都長達一整個學期！所以，在此結合工程類學生對於工業產品設計領域的已知、未知以及須知，本人盡力從眾多先進的精彩論述

1

中，結合自己多年來業界工作的心得，歸納了這樣一本在一或兩個學期內能夠完成的教材，以解燃眉之急。經幾年來「工業產品設計」課程教學，得到良好的學生反饋和實際效果。雖然不敢奢望從我們的課堂中走出許多未來的設計大師，但是希望工程類學生們經過一些基本的訓練以後，日後能夠更加得心應手地融入日益競爭的職場之中，並為臺灣的工業產品設計及開發貢獻力量，早日實現臺灣產業的轉型。本書之重頭戲在於融入了不少比較複雜的實作範例並加以詳盡敘述，試圖使讀者嘗試無師自通的感覺，也會使已經具備一些 3D 繪圖基礎技能的讀者更會感到如虎添翼。此外本書還盡量以中英文雙語進行論述，以適應全球化和外籍生日益增加的需求。

　　在此要特別感謝崑山科技大學工學院長官的卓識遠見，鼓勵筆者在機械系開設這樣的課程，此課程已經開啟十年有餘，效果頗佳。

　　由於編者水準有限，時間倉促，書中難免有錯誤和疏漏，敬請各位先進和讀者給予批評指正，不勝感激！

作者，2020 年夏於臺南

Authors, summer 2020, Tainan

目　錄

1

概述 Basic Concepts

2

工業產品設計三要素綜述 Three Basics in Product Design

參考文獻 References　　　　　　　　　　　　　267

附錄 Appendix

概述 Basic Concepts

1

以下是藝術品或玩具嗎？是手工香皂！The following are only soaps.

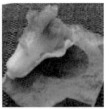

Handmade Soap from 莉詩手工皂坊 steven.chao12@msa.hinet.net

以下的戒指是白金的嗎？是不鏽鋼和水泥的組合！The rings are not made from silver or platinum, they are made from stainless steel and concrete!

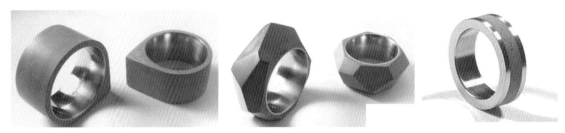

Concrete and Steel Rings From $49 to $66, 22 Design Studio, Taiwan

由此，工業產品設計所產生的附加價值可以略見一斑。These extra values are all from Industrial Product Design.

1.1　設計的定義 Definition of Design

設計的定義表達很多：

1. 所謂設計即「以假定之觀念及思維先行處理後，透過某種表述方法，訂定其擬施行之策略」通常指有目標和計劃的創作行為、活動，在藝術、建築、工程及產品開發等領域起著重要的作用。
2. 「設計是把材料經過加工，改變成新形態，產生新功能之創造性思考。」
3. 「設計是指將一個主意或計劃轉變成具有創造性的詳細的施工或生產計劃或方案。」

可是嚴格地講起來，實際上不存在一個公認的關於設計的定義，在 Wikipedia 網站上就有如下的論述：

Strictly speaking: No generally-accepted definition of "design" exists, and the term has different connotations in different fields. Informally, "a design" refers to a plan for the construction of an object (as in architectural blueprints, circuit diagrams and sewing patterns) and "to design" refers to making this plan. However, one can also design by directly constructing an object (as in pottery, cowboy coding and graphic design).

More formally, design has been defined as follows:

(noun) a specification of an object, manifested by an agent, intended to accomplish goals, in a particular environment, using a set of primitive components, satisfying a set of requirements, subject to constraints; (verb, transitive) to create a design, in an environment (where the designer operates).

摘自：Quote from: http://en.wikipedia.org/wiki/Design

首先，上帝設計創造好了自然界原始的一切，然後又締造了有願望又有能力將自然界細化及美化的人類，漸漸地在地球上進行著：市政設計、建築設計、園林設計、路橋設計、裝潢設計、服裝飾品設計以及工業產品設計，包括（交通工具、家具、電器、運動用品、科學儀器、工具機、醫療設備……）。而這些設計又是要建立在其他科學技術的發展之上的。

God created everything first and created the people who are capable of detailing and beautifying the nature including: civil; architecture; garden; decoration; cloths and ornaments; industrial products. They are all depending on science and technology.

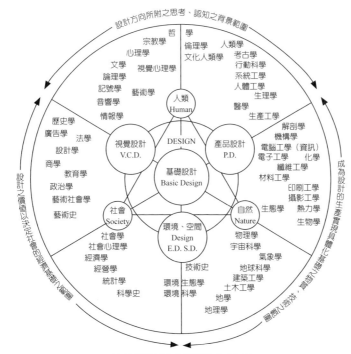

設計同其他科學緊密相關 [1]

Design depends on S&T

1.2　何謂工業產品 What Are Industrial Products

工業產品的定義表達也很多，不過筆者認為以下的表達已經夠明確了：

「因人類根據人類生活、社會的需要；設計者根據個幾的意念，激發思考的意念；應用各種天然的資源、人工的材料；配合各種加工技術，設計出來的一種產品。」工業產品是為人服務的。

There are lots of definitions of Industrial Products, the following statement are clear enough for this matter: "Industrial Products are designed, based on the needs of human beings and the society; according to the idea and creativity of the designer; utilizing natural or artificial materials; and finally with feasible manufacturing technologies."

The center of the Industrial Products is human beings.

中心是人，人是中心 [1]

The center is people

1.3　工業產品分類 Categories of Industrial Products

臺灣經濟部工業產品分類是配合民國四十二年創辦工業生產調查而編訂之分類，其後隨著我國經濟及技術的快速發展，工業產品不斷增多，為反映產業結構變遷、行業標準分類修訂及因應產業資訊需求的上升，至民國八十六年止已歷經十二次修訂。近幾年來，因資訊電子的發達，經濟自由化、國際化的衝擊，加上產業對外投資的興盛，使得國內產業結構轉變較之前更為迅速，是以政府不斷地根據現實情況重新檢討修訂產品分類。

不過總體上大致可以分為 Categories of Industrial Products:

1. 消費品——家具、電器、運動用品……

   Consumer products --- furniture, appliances, sport equipments…

2. 商業和服務業設備——醫療設備、電腦、保險箱……

   Business and service products --- medical instrument, computers, and safety box…

3. 資本貨物及耐用貨物——科學儀器、工具機……

   Durable product---scientific instruments, machine tools…

4. 運輸設備——汽車、船艇……

Transportation products --- cars, ships …

1.4 工業產品設計的定義
Definition of Industrial Product Design (IPD)

工業產品設計的定義也不止一個，例如：

「專門從事概念和系統之創造與開發，同時為使用者和製造商牟取利益。」

摘自：美國工業設計師協會（IDSA）

「在大量製造的前提下，對產品加以分析、創造、發展的作業。目的是希望在大量投資之前，能保證產品處於能廣為人們接受的方式，並且可以在一般水準的價格和合理的利潤下加以生產。」

摘自："Industrial Design" by H.V. Doren

Industrial design is a combination of applied art and applied science, whereby the aesthetics, ergonomics and usability of products may be improved for marketability and production. The role of an industrial designer is to create and execute design solutions towards problems of form, usability, physical ergonomics, marketing, brand development and sales.

The term "industrial design" is often attributed to the designer Joseph Claude Sinel in 1919 (although he himself denied it in later interviews) but the discipline predates that by at least a decade. Its origins lay in the industrialization of consumer products. For instance the Deutscher Werkbund, founded in 1907 and a precursor to the Bauhaus, was a state-sponsored effort to integrate traditional crafts and industrial mass-production techniques, to put Germany on a competitive footing with England and the United States.

摘自：Quote from: http://en.wikipedia.org/wiki/Industrial_design

1.5 工業設計的要求：三大要素
Requirements and Three Components of IPD

優秀的產品設計應具有：合理性、經濟性、審美性、獨創性。

An excellent product design must be: Appropriate; Economical; Appealing and Unique.

所謂工業設計三大要素 Three Components of Industrial Design：

1. 技術的要求 Technical——工學、工學機能、設計之物或使有用之物更爲有用（useful or more useful）。

   鞋子 Shoes：保護、止滑、耐磨、防水、彈性 1。

   刀 Knife：切割能力即，銳利度、刀刃長度、重量、尺寸、形式和材料、施力之大小。

2. 人體的要求 Ergonomically sound——人體工學；使用機能。

   鞋子 Shoes：合腳、柔軟、舒適、透氣、耐髒、彈性 2。

   刀 Knife：操作時的性質即，重量、尺寸、形狀、材質、反作用力。

3. 美感的要求 Aesthetical——美學，社會機能造型因素；綜合結構、外觀、色彩和功能以構成工業設計的「美」（beautiful or more beautiful）。

   鞋子 Shoes：樣式、形體、色彩、爽心悅目、爲之驕傲。

   刀 Knife：給予人們視覺及觸覺上的喜悅感或自豪感即，尺寸、形狀、材質、色彩。

但是，有時候三大要求並不能完全滿足，常常會發生衝突，比如增加刀的尺寸可以增加砍劈能力，但會帶來操作上的困難。那就要進行協調和優化，做出最適宜的選擇。

However, most of the times the three requirements can not always be satisfied all together, some times even contradictory. For example, the size increase of a knife can improve its capacity but will make the operation more difficult if it is too heavy. Therefore, compensation has to be made to a balance.

因此，工業設計師或者產品設計師（Industrial designer or Product designer）不同於形體設計師或者視覺設計師（Stylist or Visual Content Designer）。

Therefore, Industrial designer or Product designer are different from Stylist or Visual Content Designer.

最佳的產品設計會達到這樣的境界：現有顧客稱心滿意，潛在顧客長久嚮往。

The best product design will achieve: existing customers with satisfaction and to-be customers with anticipation.

1.6　工業設計和工程設計 Industrial Product Design and Engineering Design

作爲具有工程背景的我們的長處：在於工學方面，能夠使結構合理可靠、機構新穎現實、材料恰當、製造簡單可行、成本合適。短處：人體工學和美學的基本訓練不夠強，接觸不夠多。

Engineering designs is the keys for industrial products, because any product needs its mechanisms or functions to serve the people or the society.

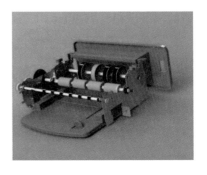

我們所熟悉的設計及專業知識：機械設計、工具設計、自動機設計、實用機構設計、產品機構設計、創意性機構設計、模具設計、機械元件設計、機械元件之系統化設計，以及機械最佳化設計。

各專業方面的知識：機械實物測繪、銲接學、機械製造、精密量測檢驗、實用板金學、塑膠模具、設計學、理論、實務、製圖、設計、微機械加工概論、鑄造學、鍛造、感測與量度工程、彈性製造系統，以及工模與夾具使用方法。

1.7　產品開發和研發
Product Research And Development

每一個企業均企盼能開發出壓倒一切競爭者的創新產品。但面對市場、技術諸多的不確定性，企業應採取正確的新產品開發程式，以及應妥善管理新產品開發的過程，才能有效的趨避新產品開發風險並創造競爭優勢。

Creative new products are the dream of any business to surpass its competitors. However because of the uncertainty of market and technology conditions, company should have appropriate new product development strategies.

一、產品開發三大部門 Three Major Departments for Product Development

1. 行銷 Sales
2. 研發 Research & Development
3. 製造 Production

二、成功產品開發五大特徵 Five common advantages for a successful product

The following common advantages are for a successful product:

1. 實用 Usefulness
2. 美觀 Beautifulness
3. 低成本或適當的成本 Low or reasonable cost

4. 易使用和易溝通 Easy to use and communicate

5. 容易維修甚至不用維修 Easy to repair even no repair

三、新產品開發的過程階段 Development for New Products

　　新產品開發由抽象意念而致具體產品。筆者認為以下敘述概括性很強，典型的產品設計過程包含四個階段：概念開發和產品規劃階段、詳細設計階段、小規模生產階段、增量生產階段。

　　A new product usually is developed from an abstract concept to its real thing. The four major steps during its development process are: Concept development and product planning; Detailed design; Small scale production; Incremental mass production.

1. 在概念開發與產品規劃階段，將有關市場機會、競爭力、技術可行性、生產需求的資訊綜合起來，確定新產品的框架。這包括新產品的概念設計、目標市場、期望性能的水準、投資需求與財務影響。在決定某一新產品是否開發之前，企業還可以用小規模實驗對概念、觀點進行驗證。實驗可包括樣品製作和徵求潛在顧客意見。

2. 詳細設計階段，一旦方案通過，新產品專案便轉入詳細設計階段。該階段基本活動是產品原型的設計與構造，以及商業生產中的使用的工具與設備的開發。詳細產品工程的核心是「設計——建立——測試」迴圈。所需的產品與過程都要在概念上定義，而且體現於產品原型中（可在電腦中或以物質實體形式存在），接著應進行對產品的類比使用測試，如果原形不能體現期望性能特徵，工程師則應尋求設計改進以彌補這一差異，重複進行「設計——建立——測試」迴圈。詳細產品工程階段結束，以產品的最終設計達到規定的技術要求並簽字認可作為標誌。

3. 小規模生產的階段，在該階段中，在生產設備上加工與測試的單個零件已裝配在一起，並作為一個系統在工廠內接受測試。在小規模生產中，應生產一定數量的產品，也應當測試新的或改進的生產過程應付商業生產的能力。正是在產品開發過程中的這一時刻，整個系統（設計、詳細設計、工具與設備、零部件、裝配順序、生產監理、操作工、技術員）組合在一起。

4. 開發的最後一個階段是增量生產。在增量生產中，開始是一個相對較低的數量水準上進行生產，當組織對自己（和供應商）連續生產能力及市場銷售產品能力的信心增強時，產量開始增加。

　　參考資料：http://www.baoku168.com/guanli/zhishi/lilun/shengchan/b1/z1/j2/020209.htm

四、現有品牌產品的研發過程 R&D for Existing Products

即使產品或產品鏈已經具有市場佔有率，企業還是要不斷的提高產品的品質。

Even for a brand name product which has stable market share and reputation, it is still necessary for the company to continuously keep and raise product quality and introduce new style even new functions.

例如：一個著名品牌鞋產品新款式開發步驟：

An Example: The development of new product lines for a named brand golf shoe company:

1. 設計師設計區域款式（美線、歐線或日本線）（100 款式 ×3 顏色 ×3 寬窄 ×1）=900 雙

   Regional product lines (American, European or Japanese)(100 styles×3 colors×3 widths×1) = 900 pairs

2. 開板、試板、結幫、貼底、樣品完成

   Samples complete

3. 行銷管理部門篩選

   Sample selections through sales department

4. 照相樣品（70 款式 ×8 顏色）=560 雙

   Picture samples (70 styles×80 colors) = 560 pairs

5. 行銷部門市場展示、反饋

   Demon by the sale department and feedback

6. 確認樣品（70×8）=560 雙

   Confirmation samples (70×8) = 560 pairs

7. 銷售樣品（70×84）=5840 雙

   Sales samples (70×84) = 5840 pairs

8. 市場確認後，確認生產試板

   Finalize production version after market confirmation

9. 下單大量生產

   Orders mass production

   週期：一年

   Cycle time: one year

五、為什麼那些工業產品設計會持久 Why These Products Last for Decades

因為它能夠綜合結構、外觀、色彩和物件功能所構成工業設計的「美」。

這裡的「美」是廣義的，就是滿足操作它的人們在視、聽、嗅、觸以及感覺上的要求。除此之外，它的生產性、品質穩定、生產價格等因素，也必須優於其他競爭產品。因此，它

的開發與優化是系統性的。即達到沒有不合理的結構、沒有機能上的缺點、不違反美的普遍妥善性的造型,因此經得起時間考驗。

Because the products combine "beautifulness" from all the aspects including: structure; form and shape; color and function.

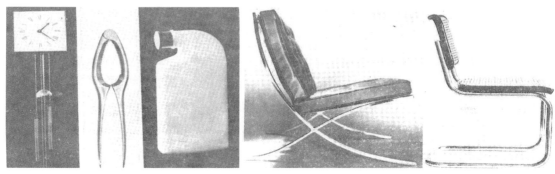

今天還能見到的德國老產品 The German products still seen today

(1)1910 年壁鐘,(2)1912 年核桃鉗,(3)1892 年漱口水瓶,

(4)1928 年巴賽隆納座椅,(5)1927 年鋼管座椅。摘自:曾坤明 [1]

1.8 定義新產品概念
Define Concepts for A New Product

明確定義新產品概念,是新產品開發過程中一件極重要的工作。新產品由於關係人利益動機的不同,一般都具有多個構面的概念。例如顧客、經銷商、供應商、研發工程師、製造人員、行銷人員之間,對於產品的認知都會有所差異。因此所謂發展核心產品概念的目的,就是希望能綜合出一個兼顧各方利益,大家都可以接受的一個產品定義。由於新產品開發往往要經過冗長的過程與各方人員的參與,因此預先形成產品開發的明確概念,將有助於增加共識與溝通,同時新產品概念也是考量企業整體競爭利益後,做出最適合的產品開發決策。

以下說明與新產品開發有關團體對於新產品概念的觀點:

Groups that contribute advices or opinions for the new product development:

1. 潛在客戶的觀點 From the to-be customers
2. 交易中間商的觀點 From the middle man dealers
3. 行銷部門的觀點 From the sales department
4. 研發部門的觀點 From research and development department
5. 生產部門的觀點 From the manufacturing department
6. 整合不同來源的觀點 Combine and integrate all the opinions

　　不同部門對所謂新產品的認知與觀點均有不同，例如：對顧客而言，它是一項能夠滿足需求的概念；行銷部門則認為，它是能迎合顧客需求的功能特色組合；對交易中間商而言，新產品必須具有競爭力，能夠創造利潤；對於研發部門而言，新產品是各種新技術運用的結果，透過系統設計與組合後產生；對於生產部門，則是新產品為零件製造與裝配的過程。因此新產品最終概念的產生，必須要結合各方的意見，經過充分的溝通與討論後形成。有時不同部門間觀點互有衝突，例如行銷部門希望新型筆記電腦價格低、重量輕體積小，但液晶顯示螢幕要大且清晰；但研發部門認為要開發大且清晰的液晶顯示螢幕需要更多的經費、時間、並且重量也要大幅增加；生產部門則希望產品設計易於裝配，不應限制產品體積的大小。為產生最終的決定，往往需要經過一番的抉擇與取捨過程，因此預先設定優先順序準則也是十分重要的。一般認為的優先順序為：滿足顧客潛在的需要、迎合中間交易商的需求、投資報酬與財務上的考慮、時間與競爭因素的考慮、本身能力上的考慮。雖然形成一個合宜、面面俱到、據創意特色、考慮周詳、充分滿足消費者需求的新產品概念，要花費許多的人力、物力與時間，但經驗證明，這是必要且值得的投入，同時也是新產品開發最關鍵的工作。

　　Different department has its own point of view or even conflict on new product development, however compensations have to be made after thorough communication and discussion.

工業產品設計三要素綜述
Three Basics in Product Design

一般來說，工業產品設計主要包括三大要素：工學、人體工學、美學。

Generally speaking, industrial product design includes three basic components: Mechanics, Ergonomics (Human Engineering) and Aesthetics.

但是隨著市場的競爭和科學技術的發展，成功的設計者必須要了解和掌握更多的知識，比如基本的材料、工藝、製程、電子、控制和通訊技術等等，才能脫穎而出並且立於不敗之地。因此，現在又提出了新的三要素定義，產品設計主要包括三個方面：功能、造型形象、物質技術基礎。功能包括物質功能（Physical Functions）和精神功能（Spiritual Functions）兩個方面：

其一，物質功能包括：

1. 技術功能：工學 Mechanics
2. 使用功能：人體工學 Ergonomics

其二，精神功能包括：

1. 審美功能：美學 Aesthetics
2. 象徵功能：形態 Style

物質技術基礎：材料、工藝、技術等各方面的製作程式中的基礎事項。對於我們工程科系的同學來說，在了解「物質技術基礎」是佔了很大的優勢。對於工學院的同學來說，我們已經經歷了不少的訓練，以滿足產品設計的「工學」要素。此外機械科同學經過專業課程，比如材料學和製造學等在了解「物質技術基礎」方面，也是佔了很大的優勢。因此本書以介紹人體工學 Ergonomics 和美學 Aesthetics 為主要內容。

2.1　工學 Mechanics and Function

Basically speaking, the mechanism or the function of a product determines its basic form.

所謂產品設計中的工學是在能夠達成同樣功能的眾多機構中，在功能至上的原則下，找出輸入與輸出的最佳匹配。因為產品的功能是產品的核心，也是產品最基本的屬性，產品的使用價值是這個產品存在的前提和價值所在，因此**產品的使用功能決定了產品形態的基本構成**。比如鞋子來說，不管是皮鞋、布鞋、跑鞋、拖鞋、膠鞋都是為了一個基本功能：「護腳」。所以不管以下圖示的鞋子是怎樣的創意、形態、形狀，仍然脫離不了「護腳」的基本功能。

Convenient shoes

可伸縮兒童球鞋

Sport shoes

運動鞋

Regular High Heels

普通高跟鞋

High Heels Greek style

希臘式高跟鞋

LadyGaGa's

女神卡卡的高跟鞋

Victoria's

辣妹的高跟鞋

Dancers'

舞女的高跟鞋

隨著人類文明和科學技術的發展，產品的功能越來越廣泛，越來越朝人性化、情感化發展。注重產品人性化的設計家 Patrick Jordan 就將產品能夠給人們帶來的快樂分成了四類：

Product brings happiness to people in four ways：

1. 生理快樂：視覺、聽覺、味覺以及觸覺。

   Physical happiness：Vision, hearing, tastes and touch.

2. 社交快樂：電話、電子郵件、明信片、賀卡等等。

   Social happiness: Telephone, e-mail, postcard and so on.

3. 精神快樂：比如擁有 iPod 或者擁有 Benz Mercedes 汽車時的那種自豪和滿足感。

   Mind happiness: The self satisfaction and proud to be a owner of an iPod or a Benz Mercedes.

4. 思想快樂：這是更高一層次的快樂，因為優質產品改善生活，尊重環境後產生的。

   Spiritual happiness: This is a higher level of happiness, such as the feeling of improved life and respect of environment due to the product.

2.2　人體工學 Ergonomics or Human Engineering

　　真正的大規模研究是受到二次大戰中對複雜設備操作的迫切需要，隨著解剖學、生態學和心理學的發展，才逐漸被重視並影響了今日的工業界。

　　It was not until World War II Ergonomics Research got much of attention, because of the urgent need of more and more complex machineries, and the advances in Anatomy, bionomics as well as psychology.

一、人體工學是人類工作和使用環境的科學研究，就是照顧人的工學 Ergonomics is the scientific research about the work and application environment

　　一個優良的產品，其簡單的人體工學要求是：

　　The simple ergonomic requirements of a good product are:

1. 容易操作使用 Easy to operate
2. 容易維護維修 Easy to repair
3. 操作使用安全可靠 Safe to operate
4. 人機介面清晰明瞭 Clear user interface
5. 人機介面新穎 Novel and interesting user interface

　　工業設計師在人體工學上的責任首先是要考慮使機器適合於人，那麼他就必須先了解人的尺度、力度和感官能力。

Country/Region	Average male height	Average female height	Sample population / age range	Methodology	Year
Indonesia	158.0 cm (5' 2.2")	147.0 cm (4' 10.0")	50+	Self-reported	1997
India	161.2 cm (5' 3.46")	152.1 cm (4' 11.88")	Rural, 17+	Measured	2007
Vietnam	162.1 cm (5' 3.7")	152.2 cm (5' 0")	25-29	Measured	1992–1993
Indonesia, East Bali	162.4 cm (5' 3.9")	151.3 cm (4' 11.5")	19–23	Measured	1995
Philippines	163.5 cm (5' 4.3")	151.8 cm (4' 11.8")	20–39	Measured	2003
Nigeria	163.8 cm (5' 4.5")	157.8 cm (5' 2.1")	25–74	Measured	1994–1996
Peru	164 cm (5' 4.6")	151 cm (4' 11.5")	20+	Measured	2005
India	164.5 cm (5' 4.8")	152.0 cm (4' 11.9")	20	Measured	2005–2006
Malaysia	164.7 cm (5' 4.8")	153.3 cm (5' 0.4")	20+	Measured	1996
China (PRC)	164.8 cm (5' 4.9")	154.5 cm (5' 0.8")	30–65	Measured	1997
Bahrain	165.1 cm (5' 5")	154.2 cm (5' 1")	19+	Measured	2002
Iraq	165.4 cm (5' 5.1")	155.8 cm (5' 1.3")	18–44	Measured	1999–2000
Korea, North	165.6 cm (5' 5.2")	154.9 cm (5' 1.0")	20-39	measured	2005
Malawi	166 cm (5' 5.3")	155 cm (5' 1.1")	Urban, 16–60	Measured	2000
China (PRC)	166.3 cm (5' 5.5")	157.0 cm (5' 2")	Rural, 17	Measured	2002
Mexico, State of Morelos	167 cm (5' 5.7")	155 cm (5' 1.1")	Adults	Self-reported	1998
Thailand	167.5 cm (5' 5.9")	157.3 cm (5' 1.9")	STOU university student	Self-reported	1991–1995
Gambia	168.0 cm (5' 6.1")	157.8 cm (5' 2.2")	Rural, 21–49	Measured	
Mongolia	168.4 cm (5' 6.3")	157.7 cm (5' 2.1")	25-34	Measured	2006

人的尺度 human dimension

　　From point of view in ergonomics, the foremost responsibility of a product designer is to make the mechanism suitable for people to operate. Therefore, he must understand first: The size, the strength and the sensibility of people.

二、設計心理學 Psychology in Product Design

　　臺灣已經快速地在國際市場上進軍，要更進一步增加臺灣產品在國際市場上的影響力，很重

汽車反光鏡：消除盲點的有效工具
A reflector: to overcome blind spot

要的步驟就是要注意提高產品的安全性和易用性。一個優良的產品設計者其實是善用一些基本的思維原理，如果沒有這些原理，設計出來的東西可能導致災難。因此隨著人類文明和科學技術的發展，心理學越來越深入產品設計領域，將心理學的原理及發現應用於工業產品的設計上。

國際知名心理學家 Donald A Norman 從心理學的角度總結了產品設計心理學的原理：

Donald A Norman, a well known psychologist summarized the principle of product design psychology:

1. 應用外界的以及頭腦中的知識

   Apply all knowledge in the mind as well as outside world

2. 簡化操作的結構

   Simplify operation mechanism

3. 系統設計顯而易見

   Clear and obvious system design

4. 正確利用配對關係

   Apply relationship of pair

5. 儘量利用自然的或人為的局限

   Utilize natural and artificial restrictions

6. 充分考慮可能存在的盲點和因此產生的人為錯誤（莫非定律）

   Consider every possible blind spot and possible human error (Murphy's Law)

7. 標準化

   Standardize

——Donald A. Norman, "The Psychology of Everyday Things"

2.3　美學 Aesthetics

Aesthetics (also spelled **æsthetics** or **esthetics**) is a branch of philosophy dealing with the nature of beauty, art, and taste, and with the creation and appreciation of beauty. It is more scientifically defined as the study of sensory or sensori-emotional values, sometimes called judgments of sentiment and taste. More broadly, scholars in the field define aesthetics as "critical reflection on art, culture and nature." Aesthetics is a subdiscipline of axiology, a branch of philosophy, and is closely associated with the philosophy of art. Aesthetics studies new ways of seeing and of perceiving the world.

——from Wikipedia

美學的定義也非常廣泛：

1. 視覺上同其他產品的差異。

2. 形象、流行及以擁有產品為傲。

3. 符合美學觀點的產品設計來激勵研發團隊的其他成員。

下面四個不同的室內布置範例顯示了美學的神奇力量 The Magic of colors。

春 Spring　　　　夏 Summer　　　　秋 Autumn　　　　冬 Winter

一般來講產品的美學主要以形體、色彩、材質（表面織構）組成。

Generally speaking, form; color and material (surface texture) are three basics.

形體 Form　　　　　　材質和表面織構 Material and Surface Texture

色彩：浪漫　　　　　色彩：魅惑　　　　　色彩：華麗

Color: Romantic　　Charming and fascination　　Gorgeous and Splendid

一、造型 Formation

　　我們說在造型動作之前，設計是不可或缺的手法之一，即產品設計是決定產品形態的行為。而產品設計是科學的、合理的思考伴隨著造型的活動。設計與造型密不可分，產品的造型往往是那個產品賣點的起始點。

　　一般來說造型分為兩大主流 Main Streams of form design：

1. 實用機能造型 Formation Follows Function
   滿足人類對產品使用方便性及效率性。

2. 審美機能造型（百看不厭）Formation for Beautifulness
   具有比例、均勻、律動、調和、重點等藝術美原則，能夠滿足人類感官上的、情緒上的及心理上的美化。

　　優秀的產品設計應具有：合理性、經濟性、審美性、獨創性。完美的產品設計也許是以上兩大主流機能的綜合，因此優秀的設計者必備：

1. 審美能力（美的要素及理論）

2. 創造性

3. 表現技術

實用和審美的結合
Combination of
function and beauty

　　Perhaps a good product design is the combination of both. And good designers must have:

1. Appreciation of beauties

2. Creativity

3. Presentation skills

　　成型的幾個有用的手段包括：比例和黃金分割法、借鑑和靈感、流線型、仿生形態學以及造型繪圖透視。

　　Several Useful Tools in Form Design including: Ration and Golden Section；Referencing and Inspiration；Stream Lining；Bionics or Bio-morphology and Perspective in Sketch.

優秀造型設計所產生的附加價值是不言而喻的 Extra value added by attractive forms

二、色彩 Color

　　色彩實在太神奇、太美妙了。色彩不僅源於大自然母親，色彩又和地理、環境、歷史文化甚至人們的感官休戚相關。長久以來人們對色彩的探索以便更好地掌控和利用，不僅是物理學和生理學，更是數學上的課題。能夠從光怪陸離的色彩歸納成有序的調色盤（Color Palette）和調色球（Color Sphare），然後又到色碼表（Color Code），在漫長的求索過程中，凝聚了多少人類知識和才華的結晶。

　　Color or **colour** (see spelling differences) is the visual perceptual property corresponding in humans to the categories called *red*, *green*, *blue* and others. Color derives from the spectrum of light (distribution of light energy versus wavelength) interacting in the eye with the spectral sensitivities of the light receptors. Color categories and physical specifications of color are also associated with objects, materials, light sources, etc., based on their physical properties such as light absorption, reflection, or emission spectra. By defining a color space, colors can be identified numerically by their coordinates.

　　Several important terminologies in color study are:

- Hue: the color's direction from white, for example in a color wheel or chromaticity diagram.
- Shade: a color made darker by adding black.
- Tint: a color made lighter by adding white.
- Value, brightness, lightness, or luminosity: how light or dark a color is.

——from Wikipedia

The colors of the visible light spectrum

color	wavelength interval	frequency interval
red	~ 700–635 nm	~ 430–480 THz
orange	~ 635–590 nm	~ 480–510 THz
yellow	~ 590–560 nm	~ 510–540 THz
green	~ 560–490 nm	~ 540–610 THz
blue	~ 490–450 nm	~ 610–670 THz
violet	~ 450–400 nm	~ 670–750 THz

可見光譜的顏色 Colors of visible light

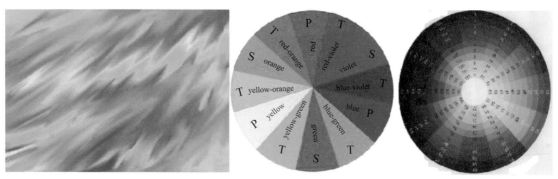

彩色圖畫與色彩盤及色彩球 Color picture, color plate and color sphere

✿ Color Code ✿

色碼選擇器　背景色彩選擇器　色碼表　英文色碼表一 英文色碼表二　爪哇色碼表

FFEBFF	FFEBF5	FFEBEB	FFF5EB	FFFFEB	F5FFEB	EBFFEB	EBFFF5	EBFFFF	EBF5FF	EBEBFF	F5EBFF
FFCDFE	FFCDE5	FFCECD	FFE7CD	FEFFCD	E5FFCD	CDFFCE	CDFFE7	CDFEFF	CDE5FF	CECDFF	E7CDFF
FFAFFE	FFAFC6	FFB0AF	FFD3AF	FEFFAF	D6FFAF	AFFFB0	AFFFD8	AFFEFF	AFD6FF		B0AFFF
FF91FE	FF91C7	FF9291	FFC991	FEFF91	C7FF91	91FF92	91FFC9	91FEFF	91C7FF		C991FF
FF74FD	FF74C8	FF7573	FFBB73	FDFF73	B7FF73	73FF75	73FFBB	73FDFF	73B7FF		
			FFAC55	FDFF55	A8FF55	55FF57	55FFAC	55FDFF	55A5FF	5765FF	AC55FF
FF37FD	FF37A9	FF3937	FF9D37	FDFF37	99FF37	37FF39	37FF9D	37FDFF	3799FF		
FF19FC			FF8F19	FCFF19	89FF19	19FF1C	19FF8F	19FCFF	1999FF		
FA00F7			FA8000	F7FA00	7AFA00	00FA03	00FA80	00F7FA			
DC00D9	DC006B	DC0800	DC7100	D9DC00	6BDC00	00DC03	00DC71	00D9DC	006BDC		7100DC
BE00BC			BE6100	BCBE00	5DBE00	00BE02	00BE61	00BCBE	005DBE		6100BE
A0009E			A06200	9EA000	4EA000	00A002	00A052	009EA0			

3

人體工學 Ergonomics and Human engineering

3.1　人體工學概述 Basics of Ergonomics

如何讓產品的設計更適合人的生理、心理特點，讓人在使用過程當中，能夠感受到舒適和便捷，以至融洽和滿足？這樣的問題也許會讓您覺得新奇而又毫無頭緒，或者說天眞而又脫離實際，或者過於簡單和平易近人，因而是無需專門來研究和思考的問題，這個問題已發展出了一個在發達國家影響巨大的現代學科。近來有些廠商，尤其在電腦和家具產品領域，把「以人爲本」、「人體工學」的設計作爲產品的特質來重點宣傳。如何讓技術的發展圍繞人的需求來展開，產品和環境的設計如何更好地適應和滿足人類的生理和心理的特點，這就是產品設計人體工程學所研究的目標。人體工學人類工作環境的科學研究，就是照顧人的工學。

Ergonomics is the science of designing user interaction with equipment and workplaces to fit the user. Proper ergonomic design is necessary to prevent repetitive strain injuries, which can develop over time and can lead to long-term disability.

The International Ergonomics Association defines ergonomics as follows:

Ergonomics (or human factors) is the scientific discipline concerned with the understanding of interactions among humans and other elements of a system, and the profession that applies theory, principles, data and methods to design in order to optimize human well-being and overall system performance.

—— From Wikipedia

眞正的大規模研究是受到二次大戰中對複雜設備操作的迫切需要，隨著解剖學、生態學和心理學的發展，才逐漸被重視，並影響了今日的工業界。工業設計師在人體工學上的責任首先是要考慮使產品或機器適合於使用的人。那麼他就必須先了解：

1. 人的尺度 The dimension of a person；2. 人的力度 The limitation of a person；3. 人的感官能力 The limitation of sense of a person。

例如：歐盟指令 90/270/EEC 和德國工作場所顯示器使用規章規定，所有帶顯示終端的工作場所不僅需要符合安全的要求，還須符合人體工學的要求，以確保勞動者的職業健康。近年來，資訊產品的人體工學認證在國際間已成爲產品認證的一個重要項目。

人體工學是一門多學科的交叉學科，研究的核心問題是不同的作業中人、機器及環境三者間的協調，研究方法和評價手段涉及心理學、生理學、醫學、人體測量學、美學和工程技術的多個領域，研究的目的則是通過各學科知識的應用，來指導工作器具、工作方式和工作環境的設計和改造，使得作業在效率、安全、健康、舒適等幾個方面的特性得以提高。人體工學從不同的學科、不同的領域發源，又面向更廣泛領域的研究和應用，是因為人機環境問題是人類生產和生活中普遍性的問題。其發源學科和地域的不同，也引起了學科名稱長期的多樣並存，在英語中，主要有 Ergonomics（歐洲）、Human Engineering（美國）等；在漢語中，則還有「人類工效學」、「人類工程學」和「人機工程學」。

　　除了我們常見的造型設計外，人體工學實際上還包括了如按鈕的位置安排、說明文字的設計等多種方面。而概括來說，所謂人體工學，在本質上就是使工具的使用方式儘量適合人體的自然形態，這樣就可以使用工具的人在工作時，身體和精神不需要任何主動適應，從而儘量減少使用工具造成的疲勞。

3.2 依人體工學設計滑鼠的範例
An Ergonomic Design Example

　　滑鼠的人體工學設計，主要就是滑鼠的造型設計。而要研究這個問題，首先需要研究人手的自然結構。

人手的自然形態 Natural structure and form of human hand

　　人手的結構中，與滑鼠相關的部分向上包括前臂，而向下則有手腕、手掌、手指等結構。

人前臂骨骼解剖結構 Bone structure of front arm

　　前臂內部包括尺骨、橈骨等主要的骨骼，人就是依靠這兩根骨頭的交錯來完成手腕的旋轉。而手腕結構中主要是一塊腕骨，它的轉動使得人的手腕可以仰俯。

人手掌的肌肉組群解剖結構 Muscle and vessel structures of palm

　　而人的手掌則主要由兩組肌肉組成；一個是拇指屈肌和外展肌組成的肌群、一個是小指屈肌及展肌組成的肌群。在兩個肌群之間有一條溝壑，對於不同的人，這條溝的深度和寬度是不同的，而這條溝內部則是人手主要神經和血管所走的地方。

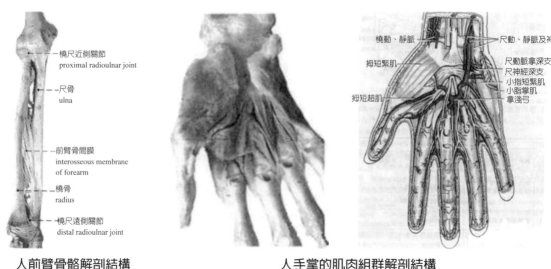

人前臂骨骼解剖結構
Bone structure of front arm

人手掌的肌肉組群解剖結構
Muscle and vessel structures of palm

　　手指的結構則相對比較簡單，每個手指包括三個指節，並在一定範圍內可以做橫向的展開。

　　這些結構的自然形態應該是什麼樣呢？

　　首先，對於上臂來說，它的自然形態應該是使尺骨和橈骨接近平行的狀態，這種狀態，也就是當前上臂和手掌平放桌上的時候，上臂和手掌呈接近垂直的傾斜狀態，使用掌外側觸及桌面的形態。因為這種形態下，上臂的主要肌肉和血管不會發生扭曲，所以即便長時間保持這個姿勢，也不會出現肌肉疲勞和缺氧情況。

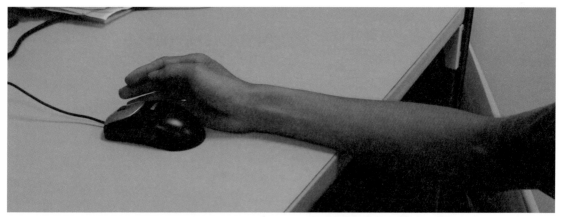

上臂的自然形態應該是使尺骨和橈骨接近平行 Natural position of the arm

　　多彩公司曾經推出過一款「豎著」使用的滑鼠，雖然由於和大多數人的使用習慣不合而沒有普及開來，但這種設計思路的確是符合人體工學要求的。

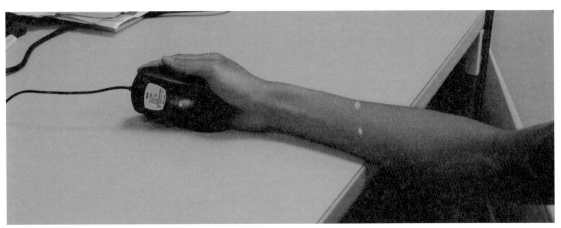

「豎著」使用滑鼠，雖然沒有普及但是符合人體工學

Vertical mouse, not popular but ergonomically sound

　　而對於手腕結構來說，多次的試驗證明，當人的手腕呈「仰起」狀態時，「仰起」的夾角在 15 度至 30 度之間的時候，是最舒適的狀態，超出這個範圍，會導致前臂肌肉處於拉伸狀態，而且也會導致血流的不暢。

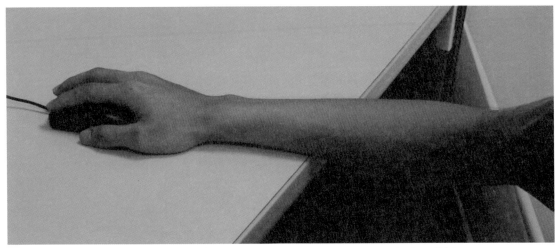

長時間扭曲會導致前臂肌肉處於拉伸狀態，且導致血流不順暢

Long time twisting will cause fatigue

　　而對於手掌來說，其最自然的形態就是半握拳狀態。而滑鼠的造型設計，實際上就是要盡量貼合這個形態。分解開來，它包括三個概念：

1. 要使滑鼠外殼緊密貼緊人手掌的兩個主要肌群：拇指肌群和小指肌群。使它們能夠貼緊而又不受壓迫。受壓迫會導致手掌處於疲勞狀態，而貼不緊又有握不住的感覺。

2. 要使滑鼠外殼緊貼掌弓而又不壓迫它。也就是滑鼠外殼要貼緊手掌中間的那條「溝」。如果它不能貼緊，那麼手心就會有「懸空」的感覺，而如果壓迫了它，下面是手主要動脈和神經的必經之地，時間長了以後會導致手缺氧。

3. 滑鼠的最高點應該位於手心而不是後部的掌淺動脈弓，否則會造成手掌產生壓迫感。

　　對於手指，手指的自然形態應該是五個手指都不懸空，而且處於呈 150 度左右的自然伸展狀態。而對於滑鼠設計來說，手指部分的一個特別要求，就是當手指自然伸展時，第三指節的指肚應該正好處於滑鼠按鍵的微動開關上，這樣才能獲得最佳的按鍵手感。

經過上述人體工學考量後設計的滑鼠造型

The mouse designs after ergonomics considerations

不良的工作姿勢就是產品設計者的機遇

Can we do some thing in design?

3.3 設計心理學 Psychology in Product Design

　　一個好的設計者其實是用到一些基本的思維原理，如果沒有善用這些原理，設計出來的東西可能導致災難。早期的廣告心理學、消費者心理學都是心理學應用的代表作品，但有關機械設計的探討，在坊間則如鳳毛麟角。臺灣已經快速地在世界市場上前進，要更進一步增加在市場上的影響力，很重要的一步就是要注意產品的易用性和安全性，因此隨著人類文明和科學技術的發展，心理學越來越深入產品設計領域，將心理學的原理及發現應用於工業產品的設計上。

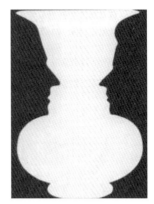

瓶子還是人影？
Vase or Faces?

　　一般來說好的產品往往綜合了以下的優點，A good product should be：

1. 容易操作使用 Easy to operate
2. 容易維護維修 Easy to maintain
3. 操作使用安全可靠 Safe to operate with high reliability
4. 人機介面清晰明瞭 Clear and intelligible user interface
5. 人機介面新穎 Clever user interface

　　可見其中大部分重點都和使用者的頭腦有關，因此作者將設計心理學歸納爲人體工學一章。Most of the requirements are connected with our mind. Therefore, "Psychology in Product Design" is arranged in this chapter.

　　國際知名心理學家 Donald A. Norman 從心理學的角度總結了產品設計心理學的原理，以使困難的工作、操作變得更容易：

Mr. Donald A. Norman, a world famous psychologist, summarized seven principles of product design psychology, for difficult operation task to be easier:

一、應用外界以及頭腦中的知識
Combine knowledge from outside and inside of the brain

　　心理學家把記憶分爲兩種：短期記憶（short-term memory）和長期記憶（long-term memory）。短期記憶是臨時，進行中的記憶，如密碼、號碼、工作中的一些暫時項目等，它的存量有限，大概只有5～7個項目，經過練習也許可以增加一倍，但這種記憶很脆弱，萬一有外界的分心干擾就會消失或被打亂，因此人所需的訊息大都存在於外界，腦中的訊息加上外界的訊息就決定了我們的行爲，所以產品設計上要考慮如何巧妙地利用這些關係，以便使用者從產品所提供的外部訊息快速地、自然地產生正確的操作反應。

該推還是拉？左邊還是右邊？　　　　只有看記號「拉」

Push or Pull? Left or Right?　　　　A sign is necessary

二、簡化操作的結構
Simplify operation mechanism

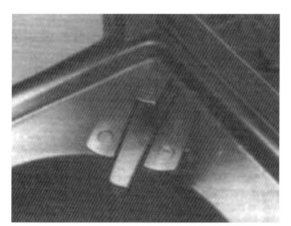

水龍頭 Water valves

三、系統設計顯而易見
System design to be obvious

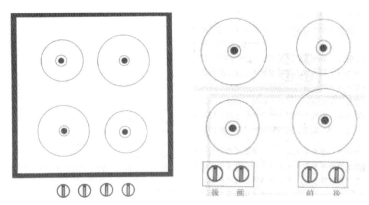

如果一個產品的操作必須依靠表明，可能它本身就是一個錯誤的設計

If an operation needs a lot of instruction, there must be mistakes in design

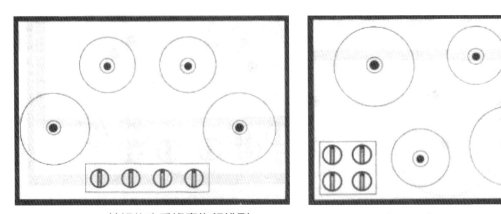

較好的廚房爐臺旋鈕排列，Better arrangement of stove buttons

四、正確利用配對關係（自然配對）
Correctly use natural match relationship

Push or Pull door design 　　　　　　　　Benz car seat control
推門或拉門不用思索　　　　　　　　　　賓士車的座椅及靠背控制設計

五、儘量利用自然或人為的局限
Utilize natural or artificial constrains

局限往往可以分為物理的局限（physical constrains）、意義上的局限（semantic constraints）、文化的局限（cultural constraints）以及邏輯的局限（logical constraints）。產品設計上如果能巧妙地利用自然或者人為的局限，使用者就能方便的知道「要怎麼做和下一步要做什麼」。A good product will fully utilizes the constrains to inform the user "What to do next".

樂高公司利用人為的物理局限巧妙地設計了童見童愛的拼裝玩具
Lego designers smartly used physical constraints in their toys

六、充分考慮可能存在的盲點和因此產生的人為錯誤
Consider possible human error and blind spots

　　你可曾有過這樣「莫非定律 *」的經驗？不帶傘時，偏偏下雨；帶了傘時，偏不下雨！在門外，電話鈴猛響；進了門，就不響了。這樣的事兒總是無可奈何，但在我們日常生活中卻是常有。「人必犯錯」，所以在設計時，設想各種使用者可能發生的錯誤至關重要。

　　"Human being will make mistakes sometime", therefore, it is key important for every possible human error or mistakes to be fully considered.

多麼聰明的車子啊，可是盧瑟（Loser）先生還是把鑰匙給鎖在車裡啦！

What a smart and intelligent design, but Mr. Loser still lock the key in the car!

七、標準化 Standardize

　　設計一個系統時標準化是非常重要，比如時鐘設計，如果順時針和逆時針的設計並行，會造成人日常生活多大的困擾是可想而知的。還有，我們在臺灣開車駕駛在左邊習慣了，如果到香港和英國就會非常困難，往往會放棄開車。

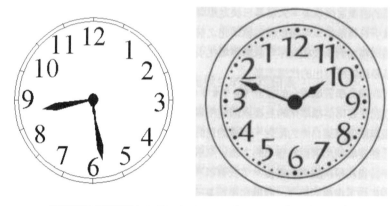

順時針或逆時針？Clockwise or Counter-Clockwise?

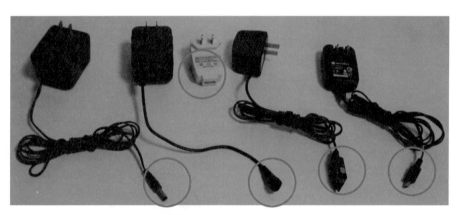

相同輸出？Same Output?

迷思供參考

Some common Myths and Contradictions (Misunderstandings) for your reference.

1. 開車越慢越安全嗎？那在高速公路上呢？

   Driving is the slower the safer, on the high-way?

2. 產品使用越方便越好？

   Designing a product, always make it easier to use.

3. 具有按鍵的產品，越少按鍵越好。

For a product with buttons, less buttons is better design.

4. 產品設計滿足 95% 的用戶就可以了？

The design is good enough for 95% of people.

5. 住「智慧屋」就沒有煩惱了嗎？

The "smart house" is "trouble free" house.

* 莫非第一定律：會出錯的事，一定出錯。

* Murphy's Law: Anything that can go wrong will go wrong, sooner than you expect, and more than once.

4

美學中的型態設計
Form and Shape Design

　　產品設計是決定產品形態的行為，而且是科學的、合理的思考伴隨著造型的活動。在造型動作前，設計是不可或缺的手法之一。

　　Product design is an action of determining the form, including scientific and logical considerations. Therefore, before any action of formation, design is indispensable.

4.1　造型分類 Categories of Form and Shape

一、自然造型 Natural Forms

非生物 No-living form

生物 Living form：植物和動物 Plants and Animals (Flora and fauna)

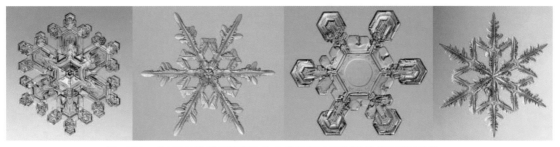

信不信，你不可能找到形狀完全相同的兩片樹葉和兩片雪花。

Believe or not: There are not leaves are identical; there are no crystals are identical.

二、人工造型 Artificial forms

人工造型和自然造型的巧妙結合 Delicate combination of artificial and natural form

人工造型和自然造型的巧妙結合 Delicate combination of artificial and natural form

人工造型兩大主流 Main Streams of artificial formation：

1. 實用機能造型 Practical Forms that follows function

   滿足人類對產品使用方便性及效率性。

2. 審美機能造型 Decorative Forms that for presentation only

   滿足人類感官、情緒及心理美化，要求百看不厭。

   一個完美的產品設計也許是以上兩大主流機能的綜合。

   A wonderful product design could combine the both of above.

實用機能造型和審美機能造型的巧妙結合 Combination of practical and decorative forms

　　立體造型的一般形式 General forms：

1. 半立體形式 Relief
2. 線立體形式 lines
3. 面立體形式 Faces
4. 塊立體形式 Blocks
5. 動立體形式 Dynamic
6. 擺動立體形式 Swing
7. 回轉立體形式 Revolve
8. 不規則曲面形式 Irregular surfaces
9. 割型立體形式 Trim and cutting

面立體形式 Faces

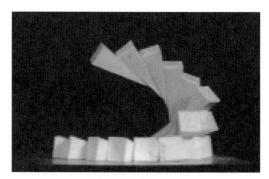

塊立體形式 Blocks

不規則拉伸和扭曲曲面　　　　　　　　　動和擺動立體形式
Surface Stretch or Twist　　　　　　　　Dynamic and Swing

造型美的形式原則 Principle of Beautifulness：

1. 反覆 Repetition

2. 比例 Ration

3. 對稱 Symmetry

4. 漸層 Gradation

5. 律動 Rhythm

6. 對比 Contrast, Conflict, Variety

7. 平衡 Balance——"the most beautiful face"

8. 調和 Harmony

9. 重點 Accent

10. 統一 Unity

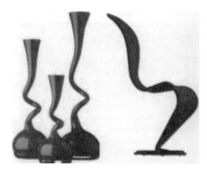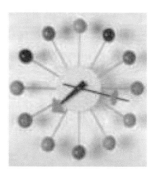

漸層 Gradation　　　　　　　　　律動 Rhythm　　　　　　　　　調和 Harmony

對比 Contrast　　　　　　　　　　　　重點 Accent

事實上我們有多少項目是可以對比（Contrast）的呢？

Please list possible pairs of contrast:

形態直接有關 Directly links to form：

直—曲（Straight-Curvature）、大—小、高—底、粗—細、長—短、寬—窄、尖—鈍、圓—角、凸—凹、遠—近、厚—薄、淺—深、粗糙—光滑、水準—垂直、平行—交叉、多—少。

和顏色或比較廣泛意義上 Relate to color or more extensive way：

硬—軟（Hard-soft）、強—弱、重—輕、慢—快、寒—暖、濃—淡、明—暗、現代—複古……眞是不勝枚舉。

4.2　成型的幾個有用的手段
Several Useful Tools in Form Design

設計者的造型表現能力 Capability of a designer
造型過程 The process of formation

1. 想法—概念—形象 Idea-Concept-Image
2. 概念的構圖 Concept sketch
3. 構圖 Sketch
4. 組裝的構圖 Assembly sketch
5. 初步造型 Rough model
6. 初步原型製作 Mockup
7. 虛擬原型（電腦輔助設計）Virtual prototype
8. 實體原型製作 Real Prototype

一、比例和黃金分割法 Ration and Golden Section

長久以來藝術家和工匠利用正交多邊形所提供的比例系統創造了多少不朽的、協調的藝術和工程上的珍品，而不需要那些複雜的計算和測量器具。正交多邊形的例子：

Orthogons provide a system of design that for centuries has allowed artists and artisans to create consistent, harmonious themes without the need for complicated calculations or measuring devices. Examples of orthogons include:

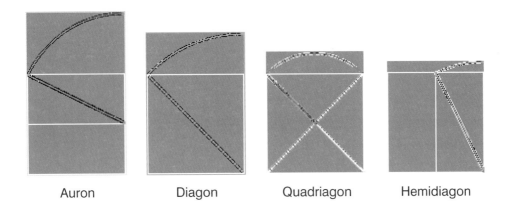

Auron	Diagon	Quadriagon	Hemidiagon

Auron: $1/2 + \sqrt{5}/2 = 1.618... = $ Phi（黃金比例 Golden Section）

Diagon: $\sqrt{2} = 1.414...$

Quadriagon: $1/2 + \sqrt{2}/2 = 1.207...$

Hemidiagon: $\sqrt{5}/2 = 1.118...$

黃金比例是一個建立在 Φ 的比例，它可以通過三角函數關係來表達：The Golden (Divine) Section is a ratio based on a phi. Phi can be related to Pi through trigonometric functions

$$2 \cdot \cos\left(\frac{\pi}{5}\right) = \phi \quad \textbf{or} \quad 2 \cdot \sin\left(\frac{\pi}{5}\right) = \sqrt{3 - \phi}$$

黃金比例的三角函數關係以及黃金比例尺

Trigonometric functions and Golden Ruler

Note: Above formulas expressed in radians, not degrees.

黃金比例通常使用希臘字母 Φ 來表達。

在黃金比例尺上，線段 A 的長短除以線段 B 的長度等於 Φ = 1.618。線段 B 的長短除以線段 C 的長度結果也一樣。

The Golden Section is also known as the Golden Mean, Golden Ratio and Divine Proportion. It is a ratio or proportion defined by the number <u>Phi</u> (Φ = 1.618033988749895...) It can be derived with a number of <u>geometric constructions</u>, each of which divides a line segment at the unique point where:

the ratio of the whole line (A) to the large segment (B) <u>is the same</u> as the ratio of the large segment (B) to the small segment (C). In other words, A is to B as B is to C. This occurs <u>only</u> where A is 1.618... times B <u>and</u> B is 1.618... times C.

—— http://goldennumber.net/

在基本代數中，黃金比例服從所謂的「Febonache 數列」1,2,3,5,8…13…。可以用微軟 Excel 作為工具來計算：

From basic algebra, it is formed by so called "Febonache Series", 1,2,3,5,8…13…. This series can be easily reproduced by Microsoft Excel as shown in the table.

i	J=J(i-1)+J(i-2)	Golden Section
1	1	
2	2	2.000 = 2/1
3	3 = 2+1	1.500 = 3/2
4	5 = 3+2	1.667 = 5/3
5	8 = 5+3	1.600 = 8/5
6	13 = 8+5	1.625 = 13/8
7	21	1.615385
8	34	1.619048
9	55	1.617647
10	89	1.618182
11	144	1.617978
12	233	1.618056
13	377	1.618026
14	610	1.618037
15	987	1.618033
16	1597	1.618034
17	2584	1.618034

實際上黃金比例被稱作為 Auron 的長方形，其中「Aur」是代表「黃金」的意思。其圖形可以用直尺和圓規在平面上實現。

The Golden Section is an Orthogon called the Auron which can be constructed from a square with a compass and ruler. This is the most commonly known of twelve orthogonses which can be constructed using this technique. Among orthogons, the golden section is known as the apron, coming from the root "aur," meaning gold.

Φ 創造美感 Phi creates a sense of beauty

在宇宙及大自然中黃金比例 Φ 幾乎無所不在，有人甚至稱之為上帝的簽字。

Phi appears throughout life and the universe. Some believe that it is the most efficient outcome, the result of natural forces. Some believe it is a universal constant of design, the signature of God.

Whatever you believe, the pervasive appearance of phi in all we see and experience creates a sense of balance, harmony and beauty in the design of all we find in nature.

It should be no surprise then that mankind would use the same proportion found in nature to achieve balance, harmony and beauty in its own creations of art, architecture, colors, design, composition, space and even music.

海螺以及許多生物成長的物理過程幾乎也是遵循著黃金比例

The physical proportions of the Seashells and many other aspects of life and the universe

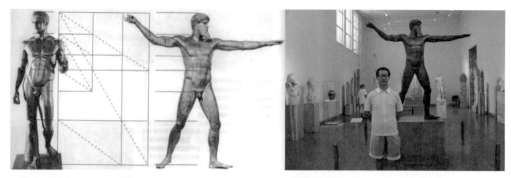

帶有虛斜線的矩形描繪了每一個黃金分割矩形，兩個希臘神話人物的比例幾乎一樣。

Bronze sculpture, thought to be either Poseidon or Zeus, National Archaeological Museum of Athens.

黃金比例的應用也許可以追溯到金字塔和馬雅神殿的設計

Its use may have started as early as with the Egyptians and Mayas in the design of the pyramids

古希臘人很早就意識並運用這個神奇的比例，賴以實現在建築及產品設計上的美觀及平衡。

The Greeks recognized it as "dividing a line in the extreme and mean ratio" and used it for beauty and balance in the design of architecture.

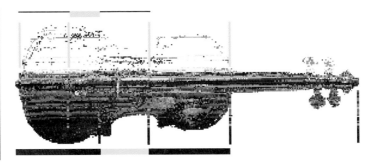

此比例被後來的人們運用於太多的藝術、建築等的設計上面。

This ratio has been used by mankind for centuries in many examples of art, architecture and design.

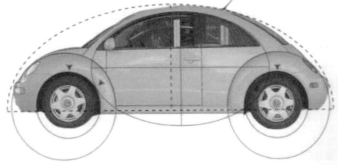

「金龜車」的設計中就體現了不少黃金比例的元素 Volkswagen Beetle and the Golden Section

是否任何世間形態都要服從黃金比例？答案當然是否定的。

However, every thing in Golden Section? The answer is no.

比如大家所熟悉的紙張設計就是運用根號 2（$\sqrt{2}$）的比例。因為這個比例的紙張可以以對折來實現等比例縮小：A2/A1＝A3/A2＝A4/A3＝A5/A4＝1/2。

For example the ratio of **Diagon**: $\sqrt{2}$ = 1.414... In the following figures, the areas A2/A1=A3/A2=A4/A3=A5/A4=1/2 Can be infinitely divided. Therefore, it is the base for DIN standard of paper dimension. Also in major European cities, it becomes the standard for billboard on the street.

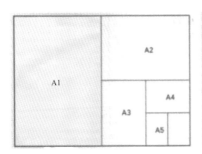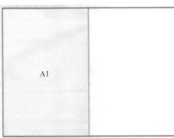

Ratio Diagon: $\sqrt{2}$ = 1.414... and the size of paper

又例如蘇格蘭聖安德魯大學精神病學院的研究人員，用電腦繪製了一張「極品男人」的臉，據說這一張臉對多數女性都可以產生無法抗拒的力量。這張合成的圖片來自 12 名學生的肖像，電腦把 12 名學生的面部特徵進行了合理的比例搭配，可以說是集 12 人之大成。

The most attractive male face synthesized by computer – "Averaging".

—— http://it.sohu.com/20060523/n243364716.shtml

二、借鑑和靈感 Reference and Inspiration

爲了取得或激發設計靈感，設計師常常使用「借鑑」這個強而有力的方法，以下便是幾項成功的案例。

To get an idea or inspiration for a design, sometimes references from other aspects of designs or objects provide powerful tools. The following pictures present several successful reference to product designs. There for a designer to be successful, it is extremely important that he/she is always looking and thinking with an open mind.

1. 以當年日本流行的女生帽子設計的檯燈 Lamp – from the idea of Japanese popular girl's hat
2. 舞動天使檯燈 Dancing Angel lamp
3. 波浪形收音機 The Wave radio from Philips
4. 炭火型煤氣爐臺 Stove design

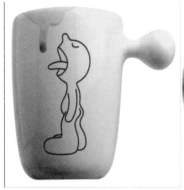

更多借鑑的例子 More examples of reference

三、流線型 Stream Line

借鑑方法另一個範例就是我們熟知的「流線型」。

One of the examples of "Reference" is "Streamline". Streamlines are a family of curves that are instantaneously tangent to the velocity vector of the flow. These show the direction a fluid element will travel in at any point in time. From basic Aerodynamics, if a traveling solid is shaped with streamlines, it will encounter the least drag force in the fluid. A **streamliner** is any vehicle that incorporates streamlining to produce a shape that provides less resistance to air.

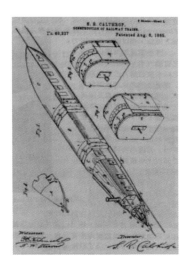

1865 年美國流線型專利。

The Calthrop Patent, 1865, #49,227 and "B&O Royal Blue" in 1937

四、仿生形態學 Bio-morphology, Bionics

另一個借鑑的範例也是我們熟知的，即「仿生學」。

Another big example of "Referencing" is "Bionics". 美國空軍軍官斯蒂爾（Jack E. Steele）少校於 1958 年首創「仿生學」此詞。

仿生學的應用實在太廣泛了。

Bionics (also known as **biomimetics**, **biognosis**, **biomimicry**, or **bionical creativity engineering**) is the application of methods and systems found in nature to the study and design of engineering systems and modern technology. Also a short form of biomechanics, the word 'bionic' is actually a portmanteau formed from *biology* (from the Greek word "βιος", pronounced "vios", meaning "life") and *electronic*. The transfer of technology between life-forms and synthetic constructs is desirable because evolutionary pressure typically forces natural systems to become highly optimized and efficient.

另一個經典的範例，就是人們從蓮花「出污泥而不染」的表面特性研發出的防水防污塗料。

A classical example is the development of dirt- and water-repellent paint (coating) from the observation that the surface of the lotus flower plant is practically unstuck for anything (the lotus effect). Examples of bionics in engineering include the hulls of boats imitating the thick skin of dolphins, sonar, radar, and medical ultrasound imaging imitating the echolocation of bats.

In the field of computer science, the study of bionics has produced cybernetics, artificial neurons, artificial neural networks, and swarm intelligence. Evolutionary computation was also motivated by bionics ideas but it took the idea further by simulating evolution in silicon and producing well-optimized solutions that had never appeared in nature.

It is estimated by Julian Vincent, professor of bio-mimetics at the University of Bath in the UK, that "at present there is only a 10% overlap".

鯊魚和飛機 Shark and airplane

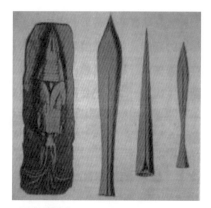

劍魚和劍 Swordfish and sword

樹和傘 Tree and Umbrella

4.3　造型繪圖透視 Perspective in Sketch

造型圖非常重要，可用來進行設計概念的擴展、修改、審視、組轉以及模擬。

Sketches are very important for Idea Expanding; Modification; Reviewing; Assembly and Simulation. For better to reflect the real objects in the sketch, perspective technique is a necessity for designers.

概念擴展 For idea expanding　　　**組裝 For assembly**　　　**模擬 For simulation**

「透視」是一種繪畫活動中的觀察方法，和研究視覺畫面空間的專業術語，透過這種方法可以歸納出視覺空間的變化規律。用筆準確地將三度空間的景物描繪到二度空間的平面上，這個過程就是「透視過程」。用這種方法可以在平面上得到相對穩定的立體特徵畫面空間，這就是「透視圖」。

Perspective (from Latin *perspicere*, to see through) in the graphic arts, such as drawing, is an approximate representation, on the flat surface (such as paper), of an image as it is perceived by the eye. The two most characteristic features of perspective are that objects are drawn:

1. Smaller as their distance from the observer increases;
2. Foreshortened: the size of an object's dimensions along the line of sight is relatively shorter than dimensions across the line of sight.

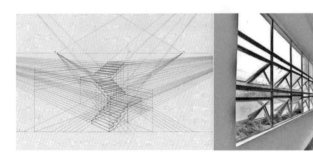

在透視概念產生以前的繪畫，圖面上人和物的大小，原則上是根據他們的重要程度來決定的，而不是距離。

Before perspective, paintings and drawings typically sized objects and characters according to their spiritual or thematic importance, not with distance, as shown in the following picture at left. Pietro Perugino's usage of perspective in this fresco at the Sistine Chapel (1481-82) helped bring the Renaissance to Rome, shown in the following picture at right.

透視法以前 Before perspective　　　　　透視法 Perspective view

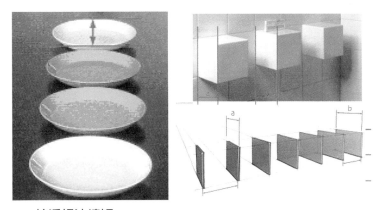

按透視法縮短 Foreshortening in perspective sketching

Horizon at eye level
地平線與眼睛平齊

Surface at eye level
平面與眼睛平齊

Extreme perspective
極限透視

從不同角度出發的透視 Perspective from different view points

4.4 電腦輔助立體造型和產品設計實例
CAD 3D Design and Practices

形與體設計練習 Possible practices in form design

1. 直稜體
2. 曲面體
3. 直稜體和曲面體

4. 切塊

5. 面的構成

6. 空間線條

7. 凸面

8. 凹面

9. 延展和發展

以下所示為本書介紹的一些電腦輔助立體造型和產品設計實例，在第八章中會詳細介紹。The following are some computer aided solid and product examples, and will be explained in Chapter 8.

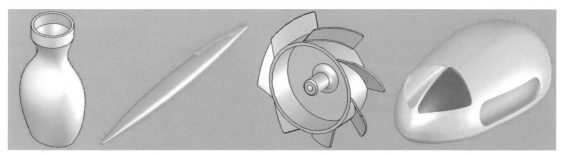

橢圓底水瓶、圓珠筆、電腦風扇以及電腦滑鼠造型設計

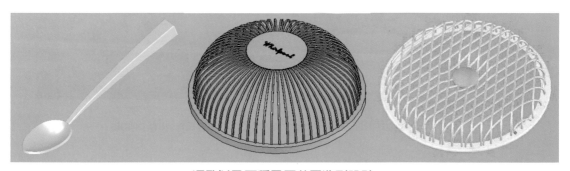

湯匙以及兩種風扇前罩造型設計

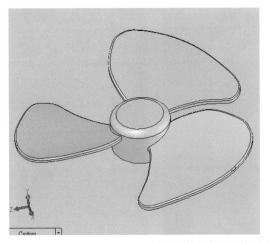

風扇葉片以及喇叭造型設計

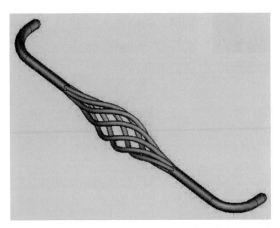
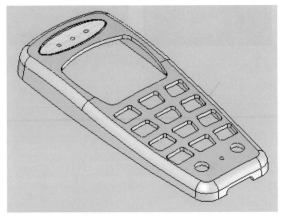

特殊扭曲螺旋形把手以及手機造型設計

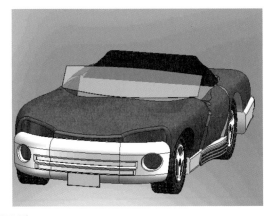

汽車外形設計

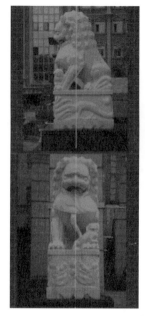

崑山科技大學圖書館前的石獅造型和 3D 數位還原

5

美學中的色彩設計
Color Design in Aesthetics

5.1　神奇美妙的色彩
Wonderful and Mysterious Colors

　　色彩實在太神奇、太美妙了。色彩不僅源於大自然母親，色彩又和地理、環境、歷史文化甚至人們的感官休戚相關。恐於言不及義，筆者只能用以下的幾組圖片來表達。

秋天的歐洲風格 Similar style of geometry, similar style of color (Fall Color Schemes)

法蘭西研究院的色彩也許是歐風的根源

Color of French Institute may be the source for European Style

幾何形態類似而色彩不同 Similar style of geometry, different style of color

幾何及表面形態相同而色彩不同

Same geometry, same texture, different color

色彩源於大自然

Colors are from Our Mother-Nature

5.2　色彩與地理和環境
Color with Geography and Environment

冰冷 Icy

熱帶風 Tropical

海洋色 Oceanic

墨西哥 Color of Mexico　　　土耳其 Color of Turkey　　　希臘 Color of Greece

5.3 色彩與歷史和文化
Color with History and Culture

西方的喜慶 Celebration in the West　　　　東方的喜慶 Celebration in the East

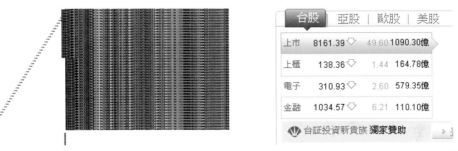

西方的股市漲 Stock up in the West　　　　東方的股市跌 Stock down in the East

西方的股市跌 Stock down in the West　　　　東方的股市漲 Stock up in the East

5.4 探索色彩 —— 物理學
Explore Color from Physics

從物理學角度可以這樣來理解，當太陽光遇到阻礙就會改變前進方向（Reflect），從而產生色彩。當光碰到物體發生折射，根據波長的不同、發生折射的幅度不同，就分化成不同的色彩。十七世紀，牛頓使用三棱鏡對太陽光線進行了分光實驗，穿過三棱鏡的太陽光，變化爲被稱作光譜的色彩帶，呈現赤橙黃綠青藍靛紫七種色彩，就跟雨後美麗的彩虹那樣。

世界上有很多色彩各異的物體，但實際上物體本身並沒有色彩。證據就是雖然在黑暗裡我們看不到蘋果的色彩，但蘋果本身確實存在。根據物體表面性質的不同，光的波長被吸收或被反射，那被反射出來的光進入我們的眼睛時，就應起了我們對色彩的感覺，色彩這才顯現在我們眼前。因此也可以說，物體的色彩是通過光的放射和吸收而呈現。

例如，紅色的蘋果將七色中的紅色波長（長波長）反射出去，將剩下的色彩都吸收了，經過反射進入我們眼睛，以紅色的形式被感知，就被稱爲物體的「表面色」。一個物體它的表面如果將所有波長的光全部反射出來，那麼我們看起來就是白色的。反之，將所有波長的光全部吸收的物體，看起來就是黑色的。世界上所有色彩都是通過光譜的七色光所產生的。

The color of an object depends on both the physics of the object in its environment and the characteristics of the perceiving eye and brain. Physically, objects can be said to have the color of the light leaving their surfaces, which normally depends on the spectrum of the incident illumination and the reflectance properties of the surface, as well as potentially on the angles of illumination and viewing. Some objects not only reflect light, but also transmit light or emit light themselves (see below), which contribute to the color also.

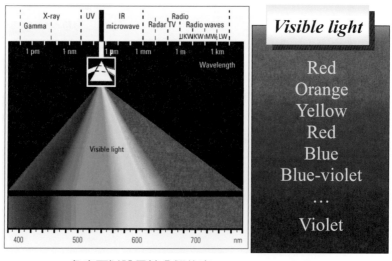

色光可以說是被分解的光 The colors of light

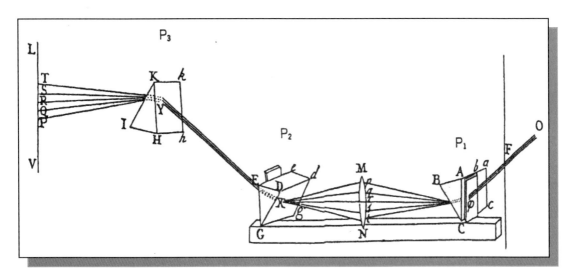

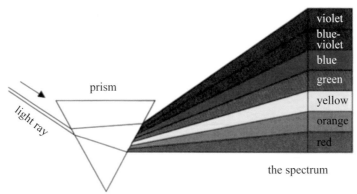

the spectrum

The visible spectrum. Of the senen lightwaves, violet has the shortest wavelength and red the longest.

牛頓（Newton,1730）混色實驗

Newton's experiment on color

5.5　探索色彩──生理學
Explore Color from Life Science

嚴格地從生理學講，光不是色彩，我們藉由人腦感知（see）色彩。

However strictly speaking, according to physiology, light is not color, it is from our sense of seeing.

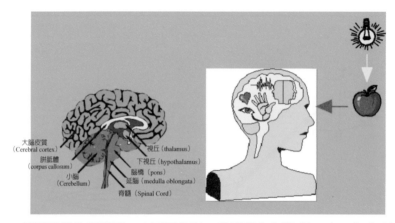

嚴格地從生理學講，光不是色彩，我們藉由人腦感知（see）色彩

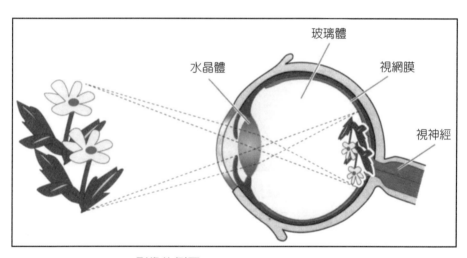

影像的倒置 Inversion of an image

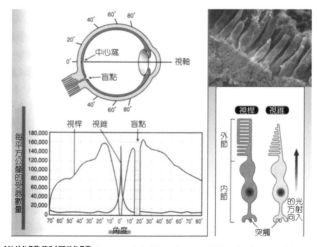

錐狀體與桿狀體 Pyramidal body and Rod-shaped body

正　常
紅色盲
綠色盲
紅—綠色盲
全色盲

色盲分類圖 Blindness chart

臺灣雲林科大學生的發明，色盲患者也可以辨識的「紅綠燈」

Invention from Taiwan Students, traffic light for Color Blindness

5.6　探索色彩 —— 數學
Explore Color from Mathematics

　　為了相互交流和應用，人們很早就開始試圖對色彩的表達進行了研究，以下僅圖列一些範例的示意。

For application and communication of color, people started color research a long time ago.

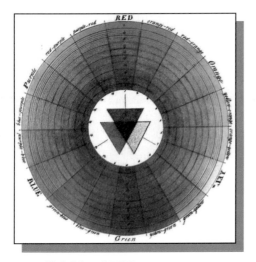
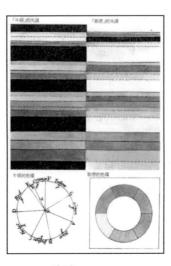
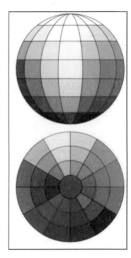

莫塞斯·哈里斯 Moses Harris
(1776)

歌德 Goethe
(1810)

菲利普·奧托·龍格
Philipp Otto Runge
(1810)

經過長久以來的研究和淘汰，才有了如今廣爲接受的色彩派和色彩球。
Color pie and Color sphere have been developed.

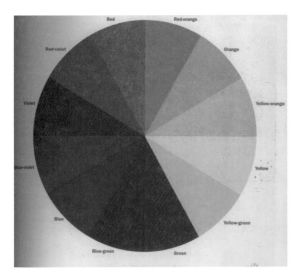
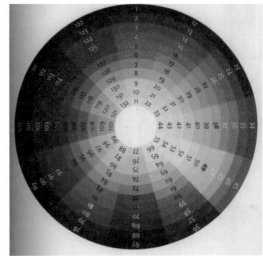

色彩派和色彩球 The Color Wheel (2D) and the Color Sphere (3D)

光的三原色及顏料的三原色 Color Light and Color Pigment

　　原色是指光線中或顏料中的色彩，無法再分解出其他的色彩，或無法以其他的色光或色料混合出來的。我們常見的色彩，大多是由兩種或以上的**顏色光**或**顏色料**所合成，三種的**原色光**分別是紅〔RED〕、綠〔GREEN〕及藍〔BLUE〕，而三種的顏料原色分別是洋紅〔MAGENTA〕、黃〔YELLOW〕及青〔CYAN〕。三原色光的混合，可得到白光；三原色料的混合，會變成黑濁色。若以適當的比例混合，則三**原色光**，或三**原色料**可調出各種不同的色彩。

Primary colors are sets of colors that can be combined to make a useful range of colors. For human applications, three primary colors are usually used. For additive combination of colors, as in overlapping projected lights or in CRT displays, the primary colors normally used are red, green, and blue. For subtractive combination of colors, as in mixing of pigments or dyes, such as in printing, the primaries normally used are cyan, magenta, and yellow, though the set of red, yellow, blue is popular among artists.

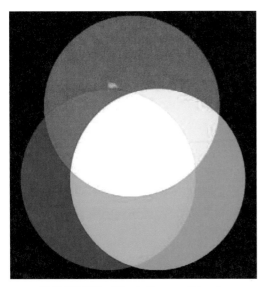

色光三原色（Red, Green, Blue）

Three basic colors in colored light

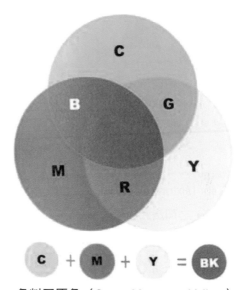

色料三原色（Cyan, Magenta, Yellow）

Three basic colors in pigment

色光三原色即一般我們所熟知的 RGB，利用光線所產生的色彩。

色光三原色若以等量的比例相加時，可獲得白色光。

色料三原色即用於印刷出版時所使用的 CMYK，C（青色，cyan）、M（洋紅，magenta）、Y（黃色，yellow）、K（黑色，black）利用顏料來產生的色彩。**色料**三原色混合後變成混濁或黑色，但不爲純黑色。

光線（色光 Colored light）的混合 mixture

將橙紅和綠的色光混合，可得到黃色光；綠和藍紫光混合可得到青綠色光；橙紅和藍紫混合可得到紅紫色光。若將三原色光混合，則會變成白光。這些色光混合後，會得到比原來色光更明亮的色光，因此色光的混合，又稱爲「**加色混合**」。

顏色 Color（色料 Pigment）的混合 mixture

　　色彩有減法，是由於物體表面上的顏料吸收了日光中一部份的光波，反射日光其他的色光。當兩種或多種顏料混合的時候，有更多的色光被吸收，越少的色光被反射，因而形成暗色或黑色。色彩的減片法是運用在顏料的混合，亦廣泛地運用在印刷技術之中，同時**色彩的減法**又稱爲 CMYK，CMYK 分別代表三原色中彩藍 C（Cyan）、洋紅 M（Magenta）、黃 Y（Yellow）以及黑 K（Black）。黑色雖然不屬於三原色的一種，但在印刷上，要加上黑顏料才能調出眞正的黑色。

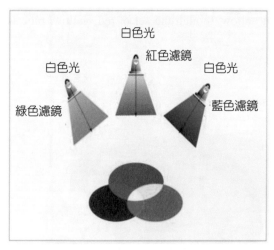

Color Mixture by addition (RGB)

加法混色原理

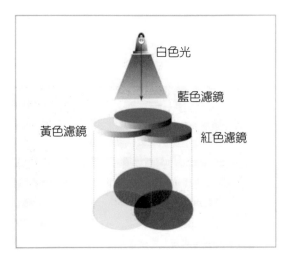

Color Mixture by subtraction (CMY)

減法混色原理

　　三原色光和三原色料之間的轉換公式 Transformation formula:

$$M + Y = R; Y + C = G; C + M = B$$

電腦顯色和三原色光數位化 Computer color digitizing

　　在電腦應用中以三原色光的 RGB 爲基礎，以下就是我們在電腦中常用的調節顏色的方法。在電腦屏幕上每一個像素（Pixel）以**三原色光** R、G、B 調製相應的顏色，而 R、G、B 分別被賦予八位二進制的記憶單元，即相當於 256 個十進制的數目。因此根據排列組合原理在一般情況下，三個八位元組成的像素可以顯示的顏色組合應該是足夠了。

　　By defining a color space, colors can be identified numerically by their coordinates, especially in digital form.

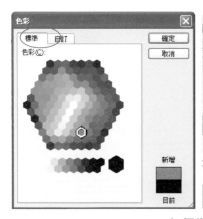
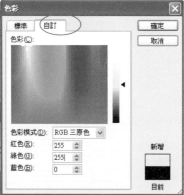

Decimal code	Binary code
00	00000000
01	00000001
02	00000010
03	00000011
04	00000100
05	00000101
...	...
10	00001010
...	...
255	11111111

每個像素的三原色共有 256 種可能的色值

Each R, G, B of Each pixel has totally 256 possible values

現在可以掌控和享受美妙的色彩了。Easily manipulate and enjoy the beautiful colors.

溫暖 Warm　　　　涼爽 Cold　　　　奇妙 Exotic　　　　光鮮亮麗 Bright

色彩繽紛 Colors close by and far beyond

6

產品設計中的創意
Creativity in Product Design

學 習 重 點

定義：創意和創新，發現和發明

定義：創意和創新，發現和發明
Creativity, Innovation, Invention, Discovery

Creativity is the ability to generate innovative ideas and manifest them from thought into reality. The process involves original thinking and then producing.

創意以非一般化的方式將看似無關的想法及觀念加以結合的能力。（陳詩雅）

創意是要超越界限，重新定義事物和事物之間的關係。也就是找出事物間的相關性，將既有的元素重新組合。（摘自〈創意學〉本書，作者為賴聲川）

Innovation is a change in the thought process for doing something, or the useful application of new inventions or discoveries. It may refer to an incremental emergent or radical/revolutionary changes in thinking, products, processes, or organizations.

中文的「**創新**」一詞是在 1980 年代才被人們接受的。英文詞語 **Innovation** 是指科技上的發明、創造，後來意義發生推廣，用於指代在人的主觀作用推動下，產生所有以前沒有的設想、技術、文化、商業或者社會方面的關係，也指自然科學的新發現。

An **invention** is a new composition, device, or process. An invention may be derived from a pre-existing model or idea, or it could be independently conceived in which case it may be a radical breakthrough. In addition, there is cultural invention, which is an innovative set of useful social behaviors adopted by people and passed on to others.[1] Inventions often extend the boundaries of human knowledge or experience. An invention that is novel and not obvious to others skilled in the same field may be able to obtain the legal protection of a patent.

發明是指一種新事物或技術的首度出現，發明可以分為物件的發明及方法的發明。而物件的發明亦不限於有形的物件、無形的物件，如電腦程式、虛擬創作等亦包括在內。

Discovery (observation), observing or finding something unknown.

發明與發現不同，發現是指一些被遺忘的事物或方法被重新發現，即使那被發現的已在世上存在過千年，甚至在人類出現以前。發現又與創新不同，創新可以是把一些舊的事物重新演譯或包裝而變成新的事物，但發明卻必須是原創的。

創意和發明就是要跳出框框，就像好的詩歌那樣：

人人心中有；人人口中無。

——From Wikipedia：摘自維基百科

—— Out of the frame.

A really jumping horse   A really running car

躍躍欲試，垂涎欲滴

6.1 創意的來源（一）：不懈的探索和嘗試 + 機遇
Sources of Creativity: Keep trying plus luck

人類歷史上偉大偶然發現 Great accidental invention or discoveries in history

Some of the biggest game-changing inventions and discoveries of our time were not the product of calculated genius, but accidents that happened to work out. These lucky mishaps have given the world everything from the awesome Slinky toy to the lifesaving antibiotic penicillin. In many cases they've also reshaped major industries or created entirely new ones. NEWSWEEK takes a look at some of the most serendipitous breakthroughs in history and how they came about

據國外媒體報導，湯瑪斯・愛迪生曾經說：「一切都需要等待，不要著急。」但是否心急只會一事無成？是不是所有進步都需要經過深思熟慮？有時天才的形成並不是靠被動選擇和等待，而是靠偶然的機遇。下面是人類歷史上十五大很偶然的發現。

1. 微波 Microwave

美國雷神公司（Raytheon）工程師珀西・斯賓塞是一位著名的電子學奇才。1945 年斯賓塞正在測試用於雷達裝備的微波輻射器（磁控管）時，突然感覺西裝褲的口袋裡有點不對勁，甚至聽到了滋滋的聲音，斯賓塞停下手裡的工作，結果發現是他口袋裡裝的一塊巧克力融化了，他猜可能是磁控管發射的微波烤化了巧克力。由此他立刻意識到也許可以把微波應用到廚房烹飪上，於是微波爐誕生了，這種廚具真可謂是全球的小吃愛好者和單身者們的大救星。

微波，發現者：珀西・斯賓塞

In 1945 Raytheon engineer Percy Spencer was testing a magnetron--a device that emits microwave radiation--when he realized that the candy bar in his pocket had melted. He figured the magnetron caused this to happen and tested his theory by placing popcorn kernels near the device. When those popped, he tried to cook an egg, which exploded. Sure, it made a mess, but he also realized that exposure to low-density microwave energy could quickly cook food. Spencer and other engineers started to work on a practical way to trap the waves and use them for this purpose. By 1947 the first commercial units became available through Raytheon. They weighed as much as 750 pounds and cost thousands of dollars, but by 1975 technological advances had made the device as popular (and affordable) as an oven range.

2. 糖精 Sweetener

糖精，發明者：伊拉・萊姆森和康斯坦丁・法赫伯格

1879 年當時正在美國約翰・霍普金斯大學實驗室工作的伊拉・萊姆森和康斯坦丁・法赫伯格回家吃飯，正吃著吃著他們突然停了下來，法赫伯格飯前忘了洗手，大部分化學家遇到這種情況都會因此身亡，法赫伯格卻意外地發現了人造甜味劑——糖精。關於這一發現，他們二人共同發表了論文，糖精的專利卻只有法赫伯格的名字，他竟偷偷申請了糖精的專利。後來萊姆森說：「法赫伯格是個無賴，讓我的名字跟他的一起出現，簡直令人作嘔。」

It may sound gross, but when Constantin Fahlberg failed to wash his hands one day in 1879, it was the luckiest thing he ever did. Fahlberg, a chemist, was at the lab of the noted scientist Ira Remsen, trying to find new uses for coal tar, when he spilled a chemical derivative on his hands. That evening, at dinner with his wife, Fahlberg noticed that his rolls tasted sweet. He asked her if she had done something special. She had not, and he quickly realized that what he was tasting was the residue he'd spilled on himself at the lab earlier that day. Eager to find out what it was, Fahlberg proceeded to taste the various residues on his arms and clothes, and later at his lab. He eventually figured out what the sweet taste was, and so did Remsen. The duo published their findings in the American Chemical Journal in 1880. Four years later, Fahlberg patented what he called saccharin--a sweetener about 300 times sweeter than sugar--but left his partner off the patent. Today saccharin is used in many low-calonies and sugar-free products, from diet soda to salad dressing.

3. 翻轉彈簧玩具 Slinky toy

翻轉彈簧玩具，發明者：理查・詹姆斯

1943 年，海軍工程師理查・詹姆斯正在想辦法用彈簧固定船上的靈敏器械，不讓它們搖晃，他隨意用手敲了敲一個原型，結果這個東西並沒摔倒在地上，而是輕輕向上彈起，然後恢復原狀。這種彈簧變成了很多人孩提時代的玩具，每年全球銷量 3 億個。

The toy was invented and developed by Naval engineer Richard James in the early 1940s and demonstrated at Gimbels department store in Philadelphia, Pennsylvania in November 1945. The toy was a hit, selling its entire inventory of 400 units in ninety minutes. James and his wife Betty formed James Industries in Philadelphia to manufacture Slinky and several related toys such as the Slinky Dog and Suzie, the Slinky Worm. In 1960, James' wife Betty became president of James Industries, and, in 1964, moved the operation to Hollidaysburg, Pennsylvania. In 1998, Betty James sold the company to Poof Products, Inc. Slinky was originally priced at $1, and has remained modestly priced throughout its history as a result of Betty James' concern about the toy's affordability for financially disadvantaged customers. Slinky has seen uses other than as a toy in the playroom: it has appeared in the classroom as a teaching tool, in wartime as a radio antenna, and in physics experiments with NASA. In 2002, Slinky became Pennsylvania's official state toy, and, in 2003, was named to the Toy Industry Association's "Century of Toys List". In its first 60 years Slinky has sold 300 million units.

4. 培樂多 Play-Doh

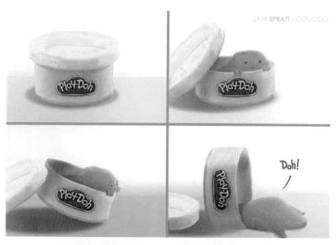

培樂多，發明者：Kutol Products 公司

　　培樂多在成為深受兒童喜愛的玩具以前，它的最初設計目的是作為清潔產品。它第一次進入市場的形象是作為骯髒壁紙的清潔物，此發現拯救了即將破產的 Kutol Products 公司，這並不是因為它的清潔效果有多麼好，而是因為小學生們開始用它製作聖誕裝飾物。該公司去掉培樂多裡的清潔劑成分，加入顏料和好聞的氣味，使它成為世界上最受歡迎的一種玩具，此一改變讓這間瀕臨破產的公司取得了巨大成功。有時候在別人沒注意到你以前，你並不清楚自己到底有多聰明。

Perhaps it's not surprising that nobody deliberately set out to invent Play-Doh, the odd-smelling, pliable goop that American kids have been shaping (and eating) for decades. In fact, Play-Doh was invented not as a toy but as wallpaper cleaner. In 1933, Cleo McVicker approached a grocery-store chain about letting his soap-manufacturing company, Kutol Products, develop a new wallpaper cleaner. The chain agreed, and Kutol started producing its cleaner, which removed the dust given off by coal furnaces. It also happened to be delightfully moldable. But demand for wallpaper cleaner eventually declined as people stopped using coal to heat their homes. Kutol Products was in peril until Cleo's son, Joseph McVicker, realized that the cleaner could double as modeling clay. In 1955 Joseph started testing the clay in schools and day-care centers--free of the cleaning agents, of course--and realized he had a hit. He named the stuff Play-Doh, and the next year he and his uncle Noah (Cleo died in 1949) created the Rainbow Crafts Co., the sole purpose of which was to make and sell Play-Doh.

5. 快乾膠 Instant Super Glue

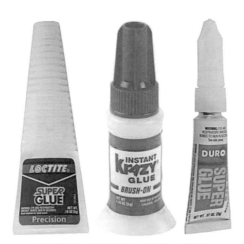

快乾膠，發明者：哈利・庫弗

1942 年，伊斯曼柯達實驗室的哈利・庫弗發現他發明的一種物質 —— 氰基丙烯酸鹽黏合劑，並不像他希望的那樣適合用在一種新的精確尺規上，因為它碰到什麼就會黏住什麼，他很快把這給忘了。6 年後庫弗在檢查為飛機駕駛艙蓋進行的一項試驗性新設計時，他再次證明氰基丙烯酸鹽黏合劑像以前一樣沒用，不過這次他注意到，這種物質不用加熱就能產生很強的黏性。庫弗和他的科研組在實驗室裡把黏性不同的物體拼接在一起，意識到他們終於為這樣東西找到了用武之地。並為這項發明申請了專利。1958 年即距離他第一次被黏住 16 年後，氰基丙烯酸鹽黏合劑開始上架銷售。

Cyanoacrylates were invented in 1942 by Dr. Harry Coover and Fred Joyner of Kodak Laboratories during experiments to make a transparent plastic suitable for gun sights.[8] Although not appropriate for the gun sights, they did find that cyanoacrylates would quickly glue together many materials with great strength. Seeing possibilities for a new adhesive, Kodak developed "Eastman #910" (later "Eastman 910") a few years later as the first true "super glue."

During the 1960s, Eastman Kodak sold cyanoacrylate to Loctite, which in turn repackaged and distributed it under a different brand name "Loctite Quick Set 404."

6. 鐵氟龍（聚四氟乙烯）Teflon

鐵氟龍（聚四氟乙烯），發現者：羅伊．普朗克特

下次你在做簡單方便的煎蛋時，一定要感謝化學家羅伊‧普朗克特，1938 年他在無意中發現了聚四氟乙烯。普朗克特本希望能生成一種新型碳氟化合物，他返回實驗室查看他在冷凍室裡進行的一項試驗，他檢查一個本應該充滿氣體的容器，結果發現氣體都已消失了，僅在容器壁上留下一些白點。普朗克特對這些神秘的化學物非常感興趣，又開始重新做實驗，最終這種新物質被證實是一種奇特的潤滑劑，熔點極高，非常適合使用在軍用設備上。現在這種物質被廣泛應用在不沾鍋上。

Roy Plunkett, a chemist at DuPont, and his assistant, Jack Rebok, were working on developing a new chlorofluorocarbon (CFC) refrigerant using tetrafluoroethylene (TFE) in 1938. The duo mixed TFE and hydrochloric acid and filled canisters they then placed on dry ice overnight. When they looked the next day, one container wouldn't open, and upon sawing the canister in half, they found that the gas had turned into a smooth, white, snowflake-like powder. Plunkett tested the new polymer and found it to be quite heat-resistant. It also had a nonstick quality. He eventually figured out how to reproduce the accidental process and patented it in 1941, then registered it under the trade name Teflon in 1944. By the 1960s, Teflon was a household name.

7. 酚醛塑料 Phenolic aldehyde Plastics

酚醛塑料，發現者：利奧‧貝克蘭德

1907 年，蟲漆被廣泛用做早期電子裝置（收音機和電話等）內部的絕緣材料，這種材料很好，不過蟲漆是用亞洲甲蟲糞便製成的，也不是電線絕緣最廉價或最簡單的方法。比利時化學家利奧‧貝克蘭德發明了世界上第一種人工合成塑膠，常被稱作酚醛塑料，這種塑膠可以壓製成各種形狀、染成各種顏色，而且在高溫環境下也不會變形且耐磨，這些優點使它成為製造商、珠寶商和工業設計者的新寵。

8. 心臟起搏器，Pacemaker

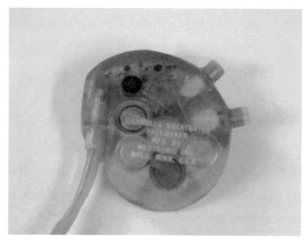

世界上第一個植入式起搏器 Photo: The world's first implantable pacemaker.

心臟起搏器，發明者：威爾森‧格雷特巴奇

　　紐約州立大學水牛城分校的副教授威爾森‧格雷特巴奇認為，他可能已經毀掉了自己的研究計畫，他不是把一個 1 萬歐姆的電阻器用在心臟記錄原型物上，而是用了 1 兆歐姆的，結果這個電路產生的信號跟人類心跳非常接近，格雷特巴奇立刻意識到，這個精確的電流也許可以調控脈搏，使因病減弱的心跳重新恢復正常。在這以前起搏器都是像電視機一樣大，是臨時性在患者身體外側使用的，現在的心臟起搏器非常小，甚至可以植入到患者的胸腔內。

　　The earliest pacemakers to regulate a heartbeat were bulky external devices that ran on AC power, meaning they--and, by extension, the patient--had to be plugged into a wall. That changed one afternoon in 1958, as Wilson Greatbatch, an engineer in Buffalo, N.Y., tried to build an oscillator to record heart sounds. Greatbatch mistakenly installed the wrong resistor in the unit, which started giving off a regular electrical pulse that matched the rhythm of a human heartbeat. Greatbatch recognized the potential of his accidental contraption and met with William Chardack, the chief of surgery at a nearby hospital, who agreed to help him. At Chardack's hospital, Greatbatch tested his device on a dog and found that the pacemaker took control of its heartbeat. "I seriously doubt if anything I ever do will ever give me the elation I felt that day when my own two cubic inch piece of electronic design controlled a living heart," he wrote in his lab diary in 1959. After another year of tinkering, Greatbatch created the world's first successful implantable pacemaker.

9. X 光 X ray

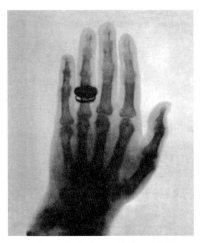

X 光，發現者：威廉・倫琴

　　X 光是自然界的一種現象，因此人類無法生成它，就連科學家發現 X 光，也非常偶然。1895 年德國物理學家威廉・倫琴正在進行一項試驗，該試驗涉及到陰極射線，當時他發現房屋對面的一張螢光紙板被點燃了，一個厚遮掩物一直放在他的陰極發射器和有輻射的硬紙板之間，這說明有光粒子穿過遮擋物，非常吃驚的倫琴很快發現，這種令人難以置信的放射物可以產生清晰圖像，第一張 X 光照片是倫琴妻子的手部骨骼。

10. 人工合成染料 Synthetic dye

　　人工合成染料則是由英國青年化學家——威廉・珀金先生，於 1856 年從一項使他失望的藥物實驗中偶爾發現的。

　　Sometimes, in pursuit of scientific endeavors, scientists become famous for reasons that don't seem entirely scientific. Take the chemist William Perkin. His big dream was to find a cure for malaria. In 1856 the young Englishman was working on an artificial form of quinine when his experiments yielded a dark sludge. It was a disappointing result, but instead of chucking the mess, Perkin noticed the color. It was a particular shade of purple, which happened to be a hot color among the fashionable folks of the time. He was able to isolate the compound that produced the color-- mauve--and realized it worked well as a dye. Within a year, Perkin patented his synthetic dye, the first synthetic dye ever made, and opened a company to make and sell it.

威廉‧珀金先生在他的實驗室（左），一位戴著紫紅色染色鏡片眼鏡的模特兒（右）
Photo: Sir William Perkin in the lab (left); a model wearing the mauve shade that Perkin created and made popular.

11. 冰棒 Popsicle

也許誰也不能想像冰棒的發明者僅是一位十一歲的孩子 Epperson（1905 年）。那天他無意中將攪拌棒留在了室外盛著飲料的杯中，因爲當時室外溫度很低，第二天飲料結成了冰。當時他不知道這個現象有什麼用處。但是，二十年後有心的他將此發現申請了專利，後來又將此專利賣給了紐約的 Joe Lowe 公司，逐漸發展到今天人人喜愛的冰棒。

Every child is grateful for Frank Epperson, even if he or she doesn't know it. In 1905, Epperson was a mere 11 years old when he accidentally left a soft-drink concoction with a stirring stick in it on the front porch of his San Francisco home. It happened to be a very cold night, and when the boy found his cup the next day, the liquid inside was frozen to the stick. While he had no idea what to make of it at the time, nearly two decades later he patented his "frozen ice on a stick" and called it the Eppsicle, but then changed the name to "Popsicle" because his children liked that better. A few years later he sold his invention to the Joe Lowe Co. in New York. Popsicle eventually ended up in the hands of Unilever's Good Humor division, which offers more than 30 flavors today.

冰棒的發明者 Frank Epperson 和他的孫女 Nancy

Photo: Popsicle inventor Frank Epperson, shown with his granddaughter Nancy.

12. 炸藥 Dynamite

阿佛烈・諾貝爾先生可謂家喻戶曉。諾貝爾發現，只要把硝化甘油和矽藻土混合在一起，加工起來就方便安全多了，1867 年，諾貝爾把這種混合配方註冊專利，並命名為「Dynamite」（結合希臘文中意為「力量」的「dynamis」和矽藻土的「diatomite」），即後來的矽藻土炸藥。1875 年，諾貝爾利用膠狀物質的火棉膠，發明出葛里炸藥（防水膠質炸藥）。

Alfred Nobel, who established the Nobel Prize, had quite the explosive history before he came up with his prestigious award. It was the 1860s, and nitroglycerin was a popular form of explosive, but it was completely unstable. That was an unfortunate problem for those trying to handle the substance, since it would unexpectedly blow up. Nobel, who owned a nitroglycerin factory, knew it was worth trying to make the compound safer, so he started to test it. One day, while researching the explosive in his lab, he accidentally dropped a vial of it on the ground. When it didn't explode, he realized it was because the substance had seeped into sawdust. The mixture essentially made nitroglycerin more stable, although not yet perfect. Nobel refined this by combining kieselguhr (a form of silica) and the explosive, making it stable enough that he could begin production of what became known as dynamite at mass scale.

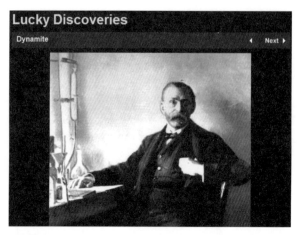

諾貝爾先生 Photo: Alfred Nobel, maker of dynamite

13. 盤尼西林（青黴素）Penicillin

盤尼西林（青黴素）是蘇格蘭細菌學家——亞歷山大・弗萊明先生於 1928 年的一場偶遇發現的。

After returning from a vacation trip in 1928, Alexander Fleming, a Scottish bacteriologist, noticed that mold had started to grow on some of the staphylococcus bacteria cultures he had left exposed. Oddly, though bacteria dotted the dish, none grew where the mold was. Fleming eventually figured out that this mold, called Penicillium notum, was causing the bacteria to undergo lysis, or membrane rupture, and killing it. It took until 1940 for Fleming's discovery to be put to practical use: that year scientists at Oxford finally isolated penicillin and developed it into the first antibiotic. Years later, while Fleming was touring a new medical laboratory, a much cleaner version than the one that had led to his famous discovery, his host commented, "If you worked here, think of what you could have invented." Fleming's response was short: "Not penicillin."

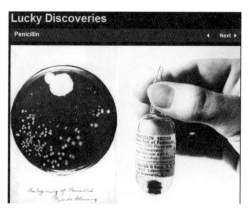

當時的培養皿（左）以及之後製成的一小瓶盤尼西林（右）

Photo: A petri dish of penicillin and staphylococci (left); a vial of the penicillin antibiotic

14. 不鏽鋼 Stainless steel

當您使用不會生鏽的刀叉時，請不要忘記英國的金屬學家——哈利‧布雷爾利先生。

The next time you raise a unrusted fork to your mouth at a meal, you should think of Harry Brearley, the English metallurgist credited with discovering the steel alloy we commonly call "stainless." Actually, stainless steel wasn't entirely Brearley's doing. Metallurgists for nearly a century before him had been toy with different metal mix, trying to create a corrosion-resistant variety. But nobody succeeded to the extent that Brearley did when he stumbled on the recipe in 1913. He had been hired by a small arms manufacturer, whose gun barrels were wearing out too quickly, to develop an alloy that would better resist erosion (not corrosion). Brearley tried elements in different proportions in the metal until he created a steel containing 12.8 percent chromiums and 0.24 percent carbon. How he figured out that his steel resisted corrosion isn't entirely verified, but the most plausible account has him running a routine test on the barrel that involved etching it with nitric acid. The metal stood up to the acid, and after it withstood other corrosives like lemon juice, Brearley realized it would be perfect for cutlery. He took his "rustless steel" to a local cutler, who dubbed it "stainless steel," and the name stuck.

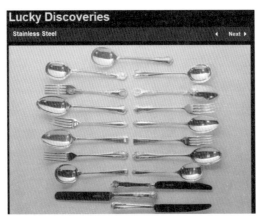

多款不鏽鋼製成的餐具 Photo: One of many cutlery sets made of stainless steel.

15. 便利貼 Post-it

雖然便利貼的發明者是 3M 公司的史賓塞‧席佛先生，但是原始的創意則始於亞瑟‧富萊先生。

While the credit for the adhesive belongs to one man, Spencer Silver, the idea for the Post-it note belongs to another: Arthur Fry. In 1968 Silver, a chemist at 3M, created a high-quality, "low-tack" adhesive, which basically means it wasn't very sticky. Silver realized it was ideal for use with paper, because the adhesive was strong enough to hold it to a surface but weak enough that paper could be removed without tearing it. The added bonus: the adhesive remained sticky through multiple uses. In seminars at the office, Silver pitched it as a surface for bulletin boards or as a spray, but after five years of trying, he continued to have difficulty finding a marketable application for it. In attendance at one of these seminars, however, was Art Fry. A colleague at 3M working in the product-development department, Fry sang in his church choir, and the paper bookmarks he used to mark his spot in the hymnal were constantly slipping out. He realized that Silver's adhesive offered a solution, and he wrote a proposal for a sticky, reusable bookmark. The samples he passed around the office were a hit, and after refining the idea, 3M introduced the Post-it nationwide in 1980.

3M 出品的便利貼 Photo: Post-it notes by 3M

6.2　創意的來源（二）：借鑑和舉一反三
Sources of Creativity:To learn by analogy

　　仿生學 Bionics：科學家所謂的「**生物仿生**」意思是靈感源自自然的人工設計。在自然進化過程中，已經形成特殊的生物鏈，可以讓萬事萬物在其中生存繁衍。下面介紹七種驚人的源自自然的「生物仿生」技術。

　　Bionics (also known as **biomimetics**, **biognosis**, **biomimicry**, or **bionical creativity engineering**) is the application of methods and systems found in nature to the study and design of engineering systems and modern technology. Also a short form of biomechanics, the word 'bionic' is actually a portmanteau formed from *biology* (from the Greek word "βιος", pronounced "vios", meaning "life") and *electronic*.

The transfer of technology between life-forms and synthetic constructs is desirable because evolutionary pressure typically forces natural systems to become highly optimized and efficient.

1. **鯊魚皮和泳衣 Sharks and swimsuit**：從電子顯微鏡下看，鯊魚皮是由稱爲「皮質鱗突」（dermal denticles）的鱗片無數重疊組成。這些鱗突在長度方向有凹槽，可以調整水在其表面的流動，這些凹槽同時可阻止漩渦或者是湍流旋渦的形成。此外，粗糙的外形還得以阻止藻類等在其身上寄生。科學家已經在泳衣設計中（現在已經在重大的比賽中禁用）和船的底部設計中利用了「皮質鱗突」特點，同時科學家還利用這種特點開發出需要阻止細菌生長的醫療技術。

2. **白蟻巢穴和辦公樓 Termite nest and office building**：白蟻巢穴看起來雖然另類，但是當外界溫度在 30 到 100 華氏度範圍變化時，非常適合居住，白蟻巢穴內溫度處於一個舒適的溫度爲 87 華氏度（這個溫度是針對白蟻來說的）。辛巴威哈拉雷伊斯特蓋特中心的建築師米克・皮爾斯（Mick Pearce）研究了白蟻巢穴涼爽的「煙囪」和「隧道」，皮爾斯將白蟻巢穴的建築理念用於 33.3 萬平方英尺的伊斯特蓋特中心建築上，使得建築比一般的建築更涼爽，而且比一般的建築節能 90%。建築物上的巨大的煙囪猶如白蟻巢穴一樣，可以在夜晚吸收涼爽的空氣用以降低樓板的溫度；而在白天樓板也可以保持涼爽，從而減少了空調的使用時間。

3. **魔鬼氈 Velcro**：瑞士工程師喬治・德・梅斯特拉爾在發明魔鬼氈的過程中，狗狗扮演了極爲重要的角色，這麼說一點也不誇張。一天梅斯特拉爾帶著他的愛犬到森林裡打獵，回來時發現狗狗身上沾了很多芒刺，稍後梅斯特拉爾在顯微鏡下觀察發現，是芒刺上的小「倒鉤」讓它結結實實地黏在織物和動物毛上，在尼龍誕生以前，他用各種織物進行了多年研究，20 年後美國國家航空暨太空總署也特別喜歡使用魔鬼氈。

Velcro is the most famous example of biomimetics. In 1948, the Swiss engineer George de Mestral was cleaning his dog of burrs picked up on a walk when he realized how the hooks of the burrs clung to the fur.

4. **鯨魚和渦輪 Whales and turbine**：生物進化讓鯨魚能夠在數百英尺處潛水幾個小時，牠們能夠獵食肉眼看不到的小型動物來維持其龐大的身材，並通過鰭和尾部來提高其運動能力。2004 年美國杜克大學、西賈斯特大學和美國海軍大學的科學家研究發現，鯨魚鰭表面的隆起可以減少 32% 的阻力。諸如 Whale Power 的公司利用這個理念，發明風力發動機渦輪葉片，大大的提高了渦輪機的效率。其他公司利用這個理念發明風扇、飛機機翼和螺旋槳。

5. **鳥群和噴氣式飛機 Birds and Jet airplane**：鳥兒採用 V 型飛行模式可以增加 70% 的飛行距離。科學家研究發現當鳥群採用熟悉的 V 形飛行時，第一隻鳥兒拍動翅膀時可以產生上升的氣流，並為後面的鳥兒所用，從而可以增加飛行距離達 70%。斯坦福大學的一組研究人員認為客運航空飛機採用同樣的策略，比單獨飛行減少 15% 的燃油量。

6. **蓮花和塗料 Lotus and paint**：蓮花有點像旱地裡的鯊魚皮。蓮花的微粗糙表面可以除掉表面的灰塵和污垢，從而保持花瓣乾淨漂亮。如果在顯微鏡下觀看蓮花，可以看到大量

的釘狀突起，這些突起可以用來抵禦灰塵。德國 Ispo 公司花費了四年時間來研究此現象，並開發出具有相似性質的塗料，具有微粗糙表面的塗料可以自動去除灰塵和污垢，從而不用洗刷外牆。

A classical example is the development of dirt- and water-repellent paint (coating) from the observation that the surface of the lotus flower plant is practically unstuck for anything (the lotus effect).

7. **甲蟲和水收集 Beetles and water collection**：沙漠甲蟲是收集水的專家，這種甲蟲由於具有獨特的外殼，可以生活在惡劣和乾燥的沙漠環境中。在沙漠甲蟲的背部有小而平整的隆起，可以充當冷凝水的收集站，甲蟲的整個背部覆蓋著光滑的蠟狀物，收集的冷凝水可以從背部流入甲蟲的口裡。麻省理工學院的研究人員從甲蟲身上獲得靈感，他們已經製作出可以更加有效的從空氣中收集水分的材料，世界上大約有 22 個國家從空氣中收集水，因此這一發明將在收集水技術方面產生重大的影響。

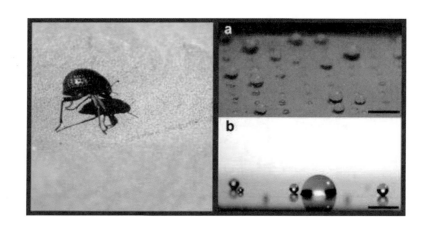

6.3 創意的來源（三）：勤於觀察，見多識廣 Sources of Creativity: Well-informed with extensive knowledge

　　以下羅列一些從網上摘錄的產品和造型創意範例以供參考，請試圖體會並解釋每一項產品或發明的用途和創意點所在。Some self explained creative examples are listed for your reference. Try to understand their application and the creativity behind them.

 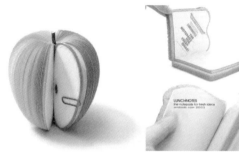

可愛的造型貼紙 Nice way of notepads and Lunch-pads

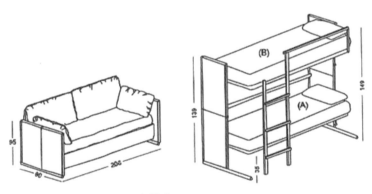

沙發床 Sofa-Bed

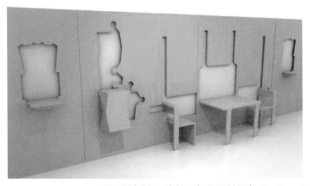 

從紙牆壁上割下來的紙板家具 Cardboard furniture from the wall

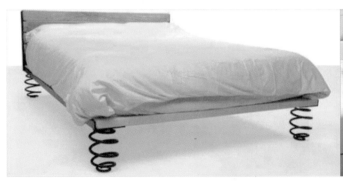

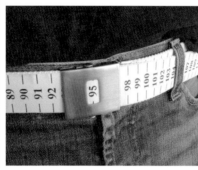

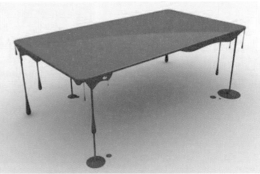

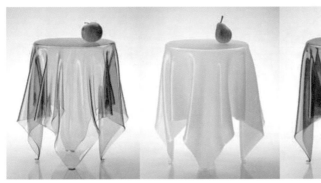

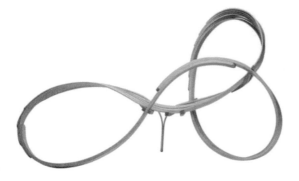

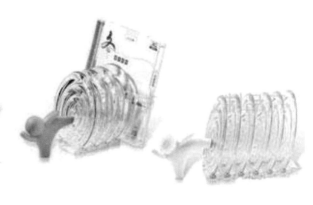

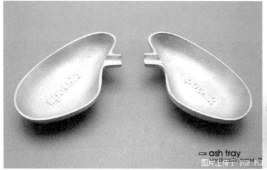

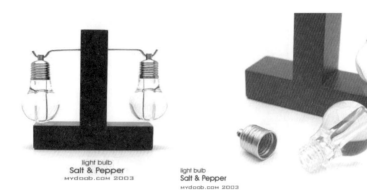

light bulb
Salt & Pepper
mydoob.com 2003

light bulb
Salt & Pepper
mydoob.com 2003

decisionmaker
mydoob 2004

Dice
주사위
mydoob 2004

Monkey
Bookmark
All we ask is that you never finish a book in one sitting.

Clip-over-the-page
Bookmark

Bookmark
mydoob.com 2003

Bookmark
mydoob.com 2003

現今時速（550 公里 / 時）最快的車 Today's fastest car - 550 km/hr（瑞士 Acabion GTBO 公司）

Before and Now Segway（賽格威）

概念交通工具 Chariot 美國 Exmovere Holdings

（賽格威／通用）及日產 Segway/GM Car and Nissan

充滿創意和探險精神的老百姓 Creative people

想飛的「汽車」Cars that want to fly

飛行汽車以及旅程 Flying Car and its trip (England)

飛行汽車 Flying Car "Transition", USA

Martin 飛行背包 Glenn Martin and his Martin Flypack 2010

英國的潛水車 sQuba (England)

挪威 Prox Dynamics 公司研製這款奈米直升機被命名為「PD-100 黑黃蜂」

新式無氣輪胎 Tires without air

6.4 創意、發明、專利和設計
Creation, Invention, Patent and Design

目前市面上有許多關於創意及設計的書籍及網站，例如：

There are lot of books and websites for explore our mind theses days. For example:

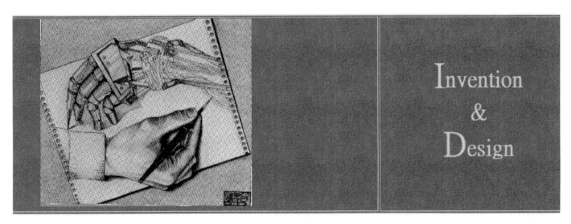

http://repo-nt.tcc.virginia.edu/inventionanddesign/home.html

供您參考：發明和設計專案 FYI: The Invention and Design Project

以上的連結是美國維吉尼亞大學研發部門的發明和設計專案網站，有興趣的讀者可以進入流覽並歡迎詢問甚至參與。The Invention and Design (I&D) project is the result of extensive research into the mental and practical aspects involved in the development of new ideas and/or products. The goal of the project is to facilitate a better understanding of the processes involved in invention and design through extended case-studies. We are dedicated to the development of the best possible materials, however, we realize the need to meet the requirements of users. In this spirit, we request that you contact us with questions, information requests, and/or comments. Principle development of the I&D resources has been accomplished by the I&D Research Team, Division of Technology, Culture and Communication, at the University of Virginia.

因此綜上所述，一個優秀、成功設計者需要：Therefore, a successful designer should：

- 必要的知識和技能 Be **Profound**, Mastering necessary knowledge and skills
- 自發的、無意識的、不由自主的 Be **Spontaneous**
- 廣泛、用心地瀏覽 Be **Aware**, Looking around with your eyes/mind
- 懂得欣賞：「喔！」、「哇！」 Be **Appreciable**, Always "Wow"
- 勤提問：為什麼？Be **Critical**, Always "Why"
- 善好奇：到底怎樣做？Be **Curious**, Always "How"

7

逆向工程和再設計
Reverse Engineering and Redesign

學 習 重 點

7.1　逆向工程的定義
Definition of Reverse Engineering

在工程和產品設計意義上講，如果把傳統的從「構思—設計—產品」這個過程稱為「正向工程」，那麼從「產品—數位模型—電腦輔助製造快速原型件」這個過程就是「逆向工程」。

但是逆向工程實際上非常廣義，在科技領域中幾乎無所不在，比如軟體的逆向工程（Decoding）、積體電路和智慧卡的逆向工程，逆向工程在軍事上的應用都有非常驚人的例子。

如果說我看得比別人更遠些，那是因為我站在巨人的肩膀上。
　　　　　　　——艾薩克・牛頓

比較嚴格和廣義的逆向工程定義：透過對某種產品的結構、功能、運作進行分析、分解、研究後，製作出功能相近，但又不完全一樣的產品過程。

　　　　　　　——維基百科

"If I have been able to see further, it was only because I stood on the shoulders of giants."
　　　　　　　—— Newton

Reverse engineering is the process of discovering the technological principles of a device, object or system through analysis of its structure, function and operation. It often involves taking something (e.g., a mechanical device, electronic component, or software program) apart and analyzing its workings in detail to be used in maintenance, or to try to make a new device or program that does the same thing without using or simply duplicating (without understanding) any part of the original.

　　　　　　　——Wikipedia

雖然逆向工程的日益發展和所謂「山寨、侵權、盜版」的質疑同時存在，但是這項技術對於科學技術的進步和普及的貢獻是無可爭議的。逆向工程可能會被誤認為是對智慧財產權的嚴重侵害，但是在實際應用上，反而可能會保護智慧財產權所有者，例如在積體電路領域，如果懷疑某公司侵犯智慧財產權，可以用逆向工程技術來尋找證據。

7.2　逆向工程在產品設計和工程上的應用
Applications of RE in Product Design

　　在機械工程或者在造型技術上，逆向工程顧名思義就是反其道而行，為結合 CAD／CAM 系統與三次元量測系統，測出數據資料以逆向軟體進行點資料處理，經過分門別類族群區隔點線面與實體誤差比對後，重新建構曲面模型、產生 CAD 資料。因此本書的「逆向工程」僅限於此類。

　　機械工程的學生都記得在基本量測課程中老師所教的「測繪」。老師拿來一個簡單的機械組合體，讓同學用量規取得長寬高以及直徑等，並且以製圖的形式還原機械組合體的設計圖，其實這就是逆向工程的一項基本訓練。可是大多數的形狀都是不規則的，例如圖示的風扇葉片和公雞模型，就不能以一些簡單的圓柱和平面等幾何元素成型的。因此就需要所謂的逆向工程技術來進行 3D 立體還原或再造、再設計。

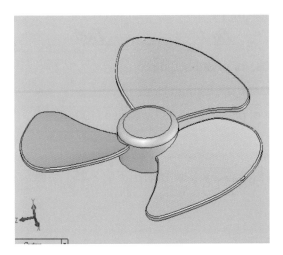 

　　因此多少年來，人們就一直致力於這項研究，甚至可以追溯到 19 世紀。以下就分別是1860 年的 3D 塑像過程和 1904 年美國專利中敘述用照相方法進行 3D 立體模型製造的圖示。

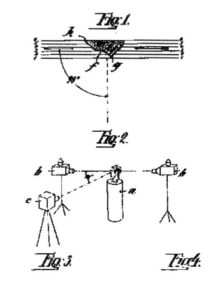

Admiral Farragut sits, late 1860s, for photo-sculpture

(Bogart 1979; photo courtesy of George Eastman House).

Photographic process for the development of plastic objects by Baese (1904)

　　逆向工程技術眞正的突破和普及是隨著電腦軟硬體的成熟發展才成爲可能，這一、二十年來更是突飛猛進，從事逆向工程的掃描硬體和編輯掃描資料的軟體層出不窮、舉不勝舉。

7.3　逆向工程的分類和方法
Categories of Reverse Engineering

　　總的來說，逆向工程掃描分兩種類型：1. 外部曲面；2. 內部結構。

1. 外部曲面 External Surfaces
　(1) 接觸掃描 Contact Scan
　　　立體掃描器 3D Digitizer
　　　CNC Probe, CMM
　(2) 非接觸掃描 Non-Contact Scan
　　　雷射掃描儀 Laser Scanner
　　　立體照相機 3D Camera

掃描得到的數據為點雲，The format of scanned is Point cloud usually in text file

X1, Y1, Z1

X2, Y2, Z2

…

Xn, Yn, Zn

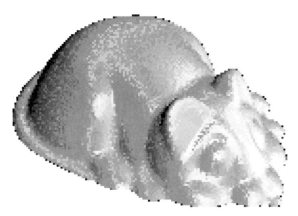

外部曲面模型和相應的點雲 External Surface Model and Corresponding Point Cloud

2. 內部結構 Internal structure

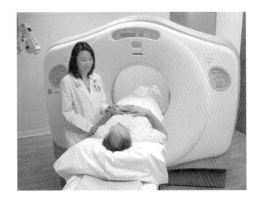

(1) 核磁共振 MRI scan

(2) 電腦斷層掃描 CT scan

(3) 超音波 Ultrasonic scan

(4) 顯微造型 Microscopy

(5) 破壞性切層 Destructive slicing

掃描得到的數據為平面影像堆疊 Plane image stacks；檔案格式一般為 file format：jpg、bmp、dicom、tiff……

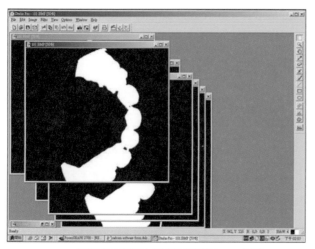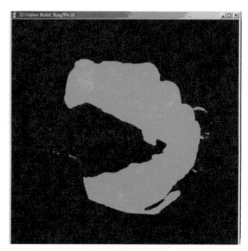

牙齒掃描照片疊層後還原 3D 立體 3D Reproduction of Teeth with Slicing

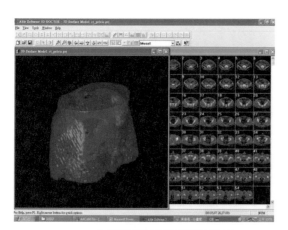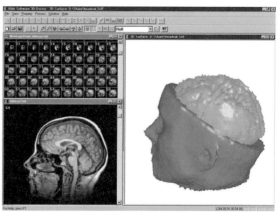

內部結構和平面影像堆疊 Internal structure and Plane Image Stacks

7.4　產品造型和外部曲面的逆向工程方法
Methods of RE on External Surfaces

　　而一般來說，產品造型設計和外部曲面有關，故本書重點談論外形的掃描和 3D 成型。外部曲面掃描器的種類繁多，不勝枚舉，以下僅列出典型的類型。

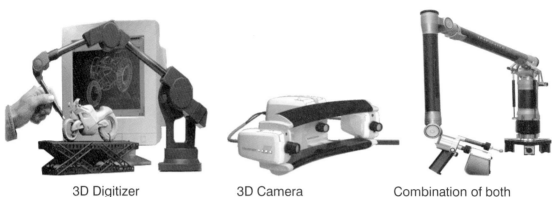

3D Digitizer	3D Camera	Combination of both
3D 點接觸掃描器	3D 照相機	3D 點接觸掃描器及 3D 照相機組合

　　一般來說，從掃描器得到的是點資料，如果被掃描的物件太大或過於複雜，則需要分幾個面進行掃描，然後利用逆向工程軟體中的對齊（Align command）指令將各個方位的點資料進行重新排列，對齊動作完成以後，一般可以進行兩個動作：

　　利用三角化指令對模型進行實體化，這樣所得到的模型以 STL 格式輸出以後可以直接進行 CAM 加工或 RP 快速成型，可是如果要對模型在 CAD 中進行進一步的編輯甚至再設計，就需要鋪曲面的動作了。以下兩組圖分別是安全帽和製作鞋楦的逆向工程過程。值得指出的是，逆向工程技術日新月異，以上所述僅是時至今日的發展狀況。

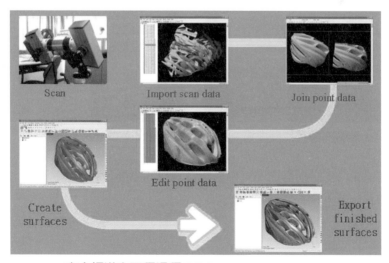

安全帽逆向工程過程 RE Process for a Helmet

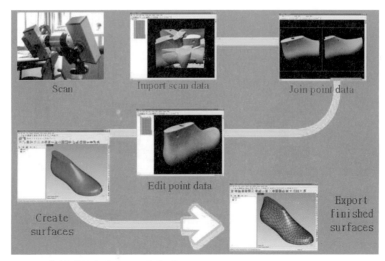

製作鞋楦逆向工程過程 RE Process for a Shoe Last

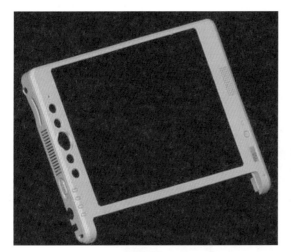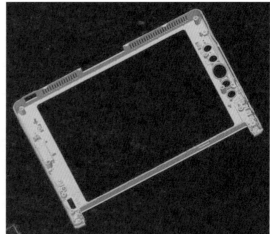

相機外殼正反面掃描 RE Scan Result of A Camera Frame

　　除了利用專業的 3D 掃描器以外，如果讀者的目的主要是在於再設計，則第八章介紹的逆向工程平面照相法不乏是一個很好的手段。

8

產品造型設計實作範例
Design Practice Examples

學 習 重 點

本章非常詳盡地介紹一些有趣及實用的產品造型以及商標設計範例，並通過造型過程使讀者掌握電腦軟體很多進階的功能，且儘量使讀者感覺到「無師自通」，照著書中步驟就可以依樣畫葫蘆地完成看似複雜的造型。In this chapter, designs of several products are to be explained in detail.

8.1　音箱造型設計 Speaker Design

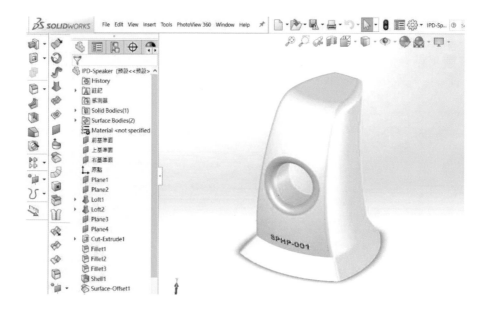

1. 開啟零件檔，顯示基本座標面，並設爲等視角（Isometric）。

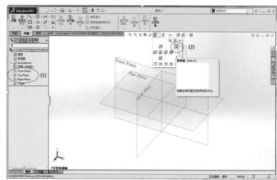

2. 建立圖示之平面 1 及平面 2。

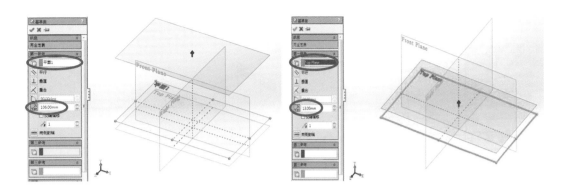

3. 在新建的平面 1 之上繪製如下草圖（四個圓弧）。

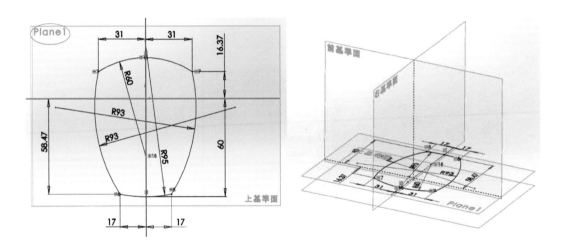

4. 在新建的平面 2 之上繪製如下草圖，捷徑之一可以利用偏移（offset）。

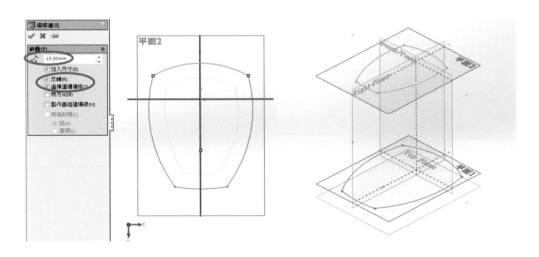

5. 在右基準面上繪製一半徑 120mm 之圓弧，並以增加關係（Add Relation）工具將圓弧兩個端點分別與剛剛繪製的草圖之圓弧相交。

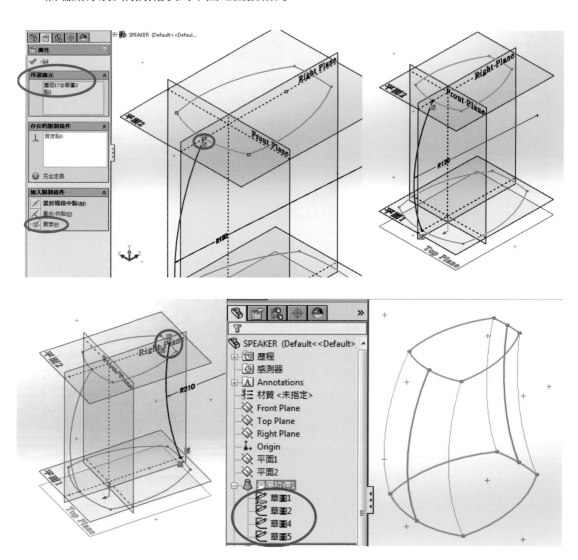

6. 以疊層拉伸（Loft）的方法，並結合半徑為 120 以及 210mm 的引導曲線（Guide curves），產生了音箱的主體實體。

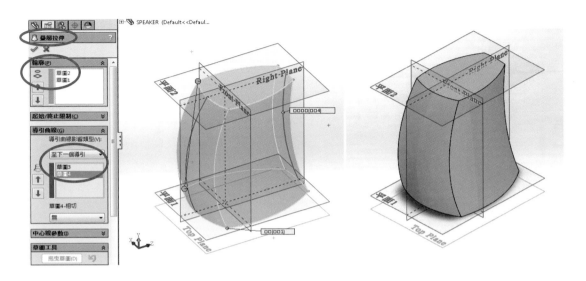

7. 在 Top Plane 上以偏移工具繪製另一草圖。Draw the following sketch on the top plane, use offset tools.

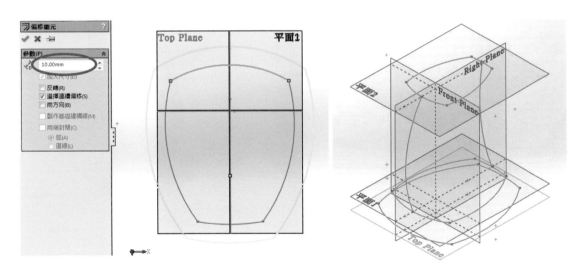

8. 運用疊層拉伸工具產生底座。Using the loft tools to create body.

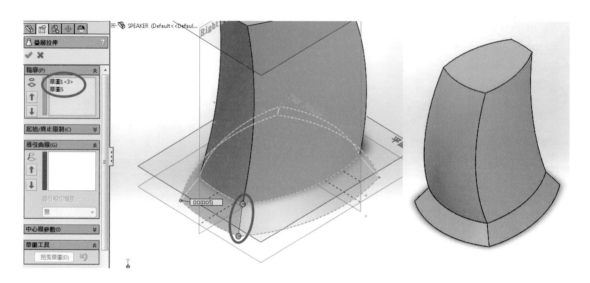

9. 建立一個如下的平面。Create the new plane using front plane.

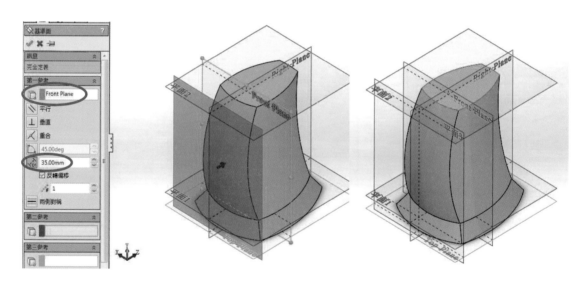

10. 再產生一個平行平面 4。Create the new plane using plane 3 which is created new plane.

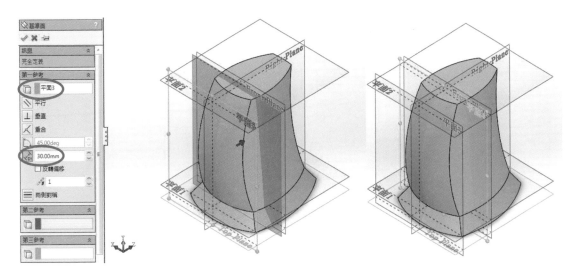

11. 在平面 3 上畫一個圓，拉伸除料至平面 4。Draw circle on plane 3 and using extruded cut tools to cut the body till plane 4.

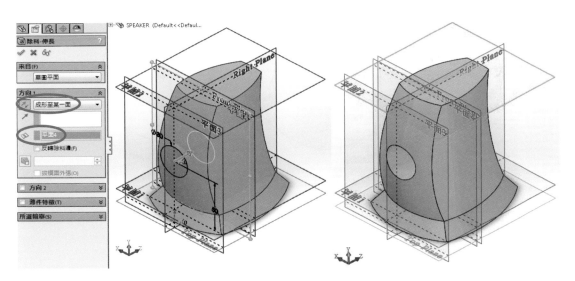

12. 產生圓角。Make it smooth using fillet tools。

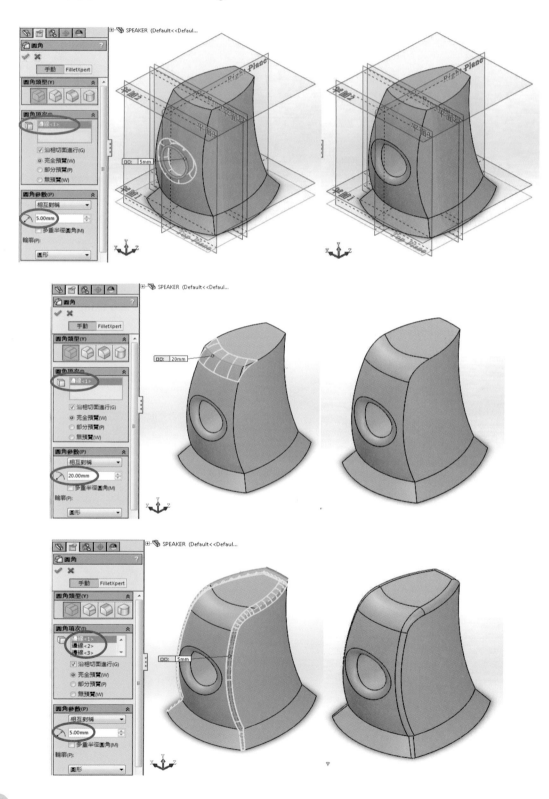

13. 產生 2mm 薄殼，完成箱體。Using shell tools for cut inside the body.

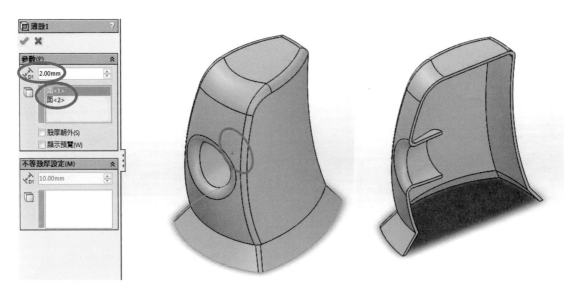

　　以下步驟可以在曲面上刻字。

14. 產生利用曲面偏移工具產生兩個曲面。Create surface using OFFSET SURFACE TOOL.

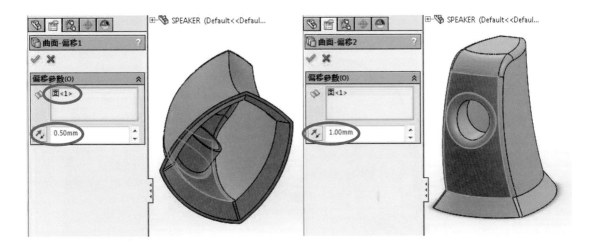

15. 在平面 4 上寫字並產生拉伸實體。Sketch on plane 4 and write what you want. Then using extruded boss and base.

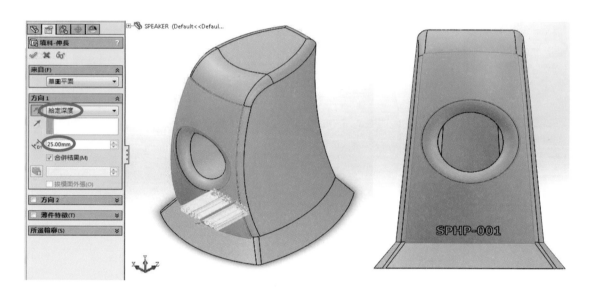

16. 使用曲面除料將多餘的部分切除。Using CUT WITH SURFACE tools to cut outside of surface.

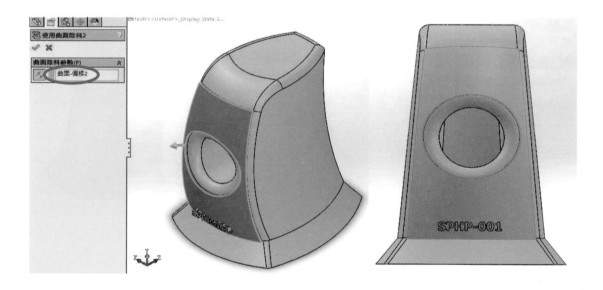

17. 使用曲面除料將多餘的部分切除。Using CUT WITH SURFACE tools to cut inside.

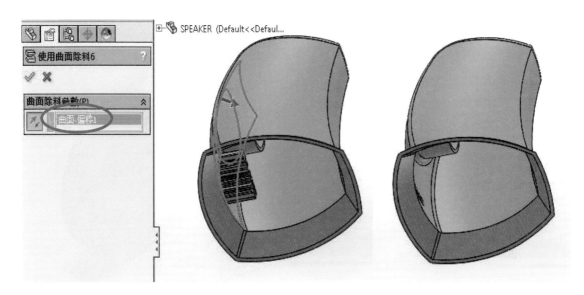

18. 將曲面隱藏並顯示出清晰的字體。Hide the surfaces.

8.2　電腦風扇設計 CPU Fan Design

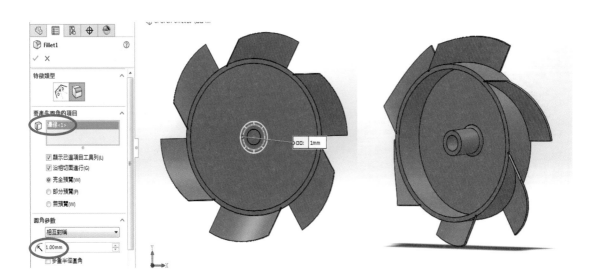

1. 在前基準面上繪製直徑為 50mm 的圓，以伸長填料產生圓柱體。Sketch on front plane and draw circle and using extruded boss/base tools to make it solid body.

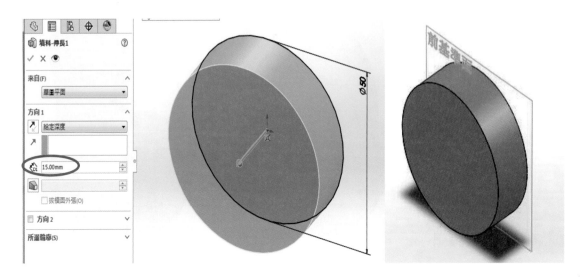

2. 產生 1.5mm 的薄殼。Using shell tool with 1.5mm thickness.

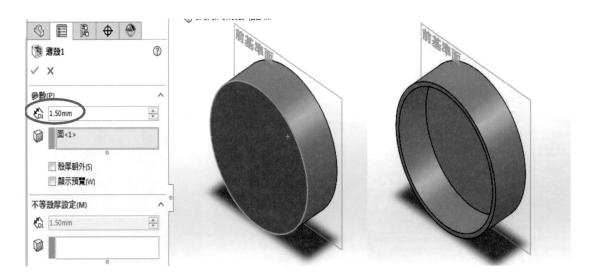

3. 在上基準面繪製如下草圖，並將此草圖投影至圓柱面。Sketch on top plane draw curve and using Project curve tool to onto body face.

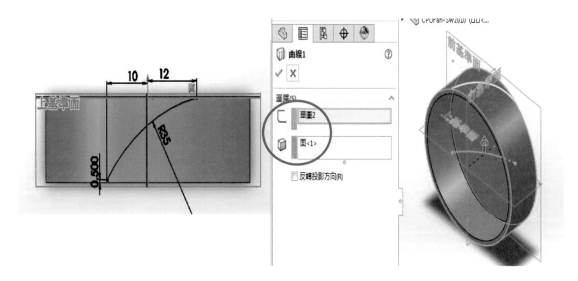

4. 在曲線端產生一平面。Create plane using curve end point which created before.

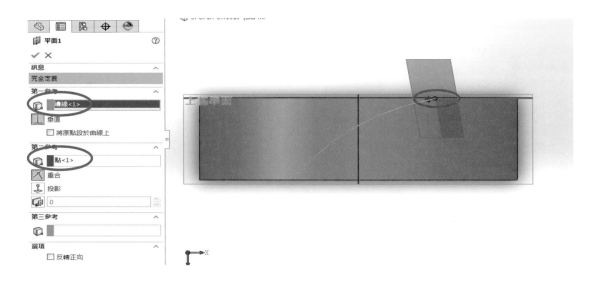

5. 在新產生的平面上繪製以下的草圖，並以掃出填料產生一葉片。Sketch on new plane draw rectangle and using swept boss/base tool to make it body.

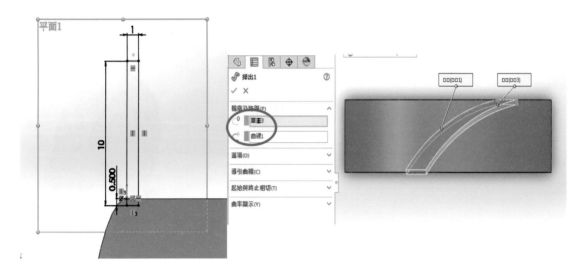

6. 利用環狀複製排列工具產生其他葉片。Using circular pattern tool to make it 6 duplicate blades.

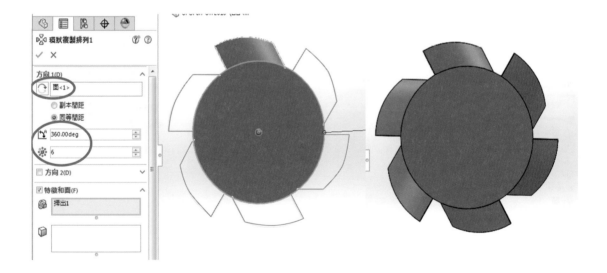

7. 產生如下的風扇軸。Sketch on inside the body face draw circles and using extruded boss/base tool.

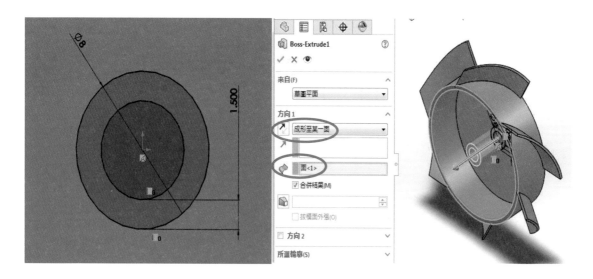

8. 產生圓角。Using fillet tool to make it smooth.

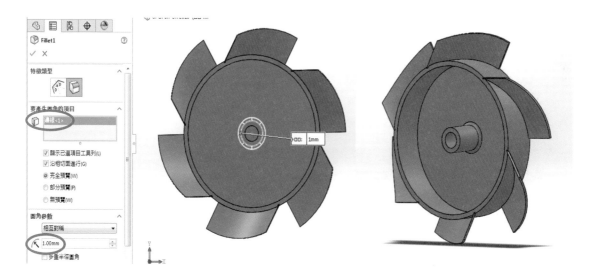

8.3　電腦鍵盤設計 Keyboard Key Design

1. 在上基準面上繪製如下長方形。Sketch on top plane draw rectangle.

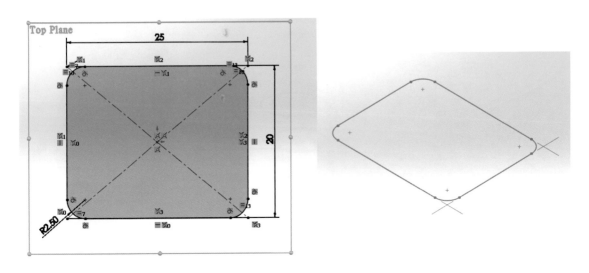

2. 在右基準面上繪製如下草圖，在中心線頂端建立一個基準面。Sketch on right plane. Draw centerlines for create new plane.

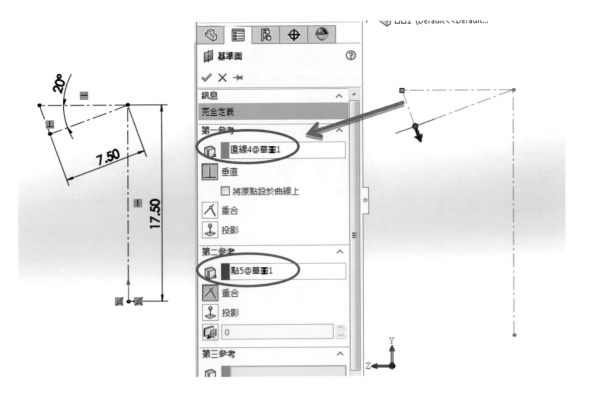

3. 在新產生的平面上繪製如下長方形。Sketch on new plane which created before and draw rectangle.

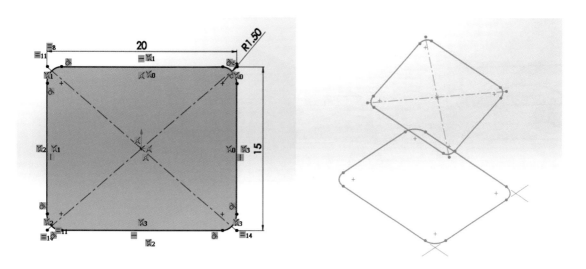

4. 運用疊層拉伸產生以下實體。Using Lofted boss/base tool to make it body.

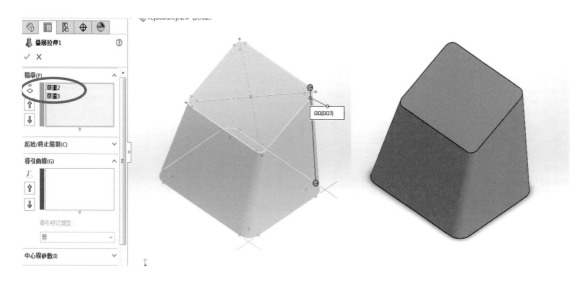

5. 在中心線頂端再產生一個平面。Create plane using centerline which created before.

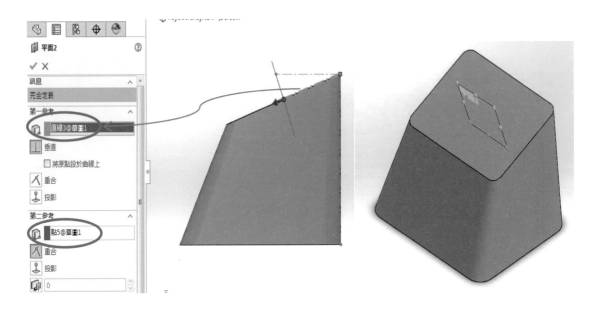

6. 在新產生的平面以及右基準面上繪製兩條曲線，曲線的端點與實體的邊緣必須利用「建立關係」功能進行連接。Create curves on the new plane and right plane. The center points of the curves should connect with line and relations should pierce.

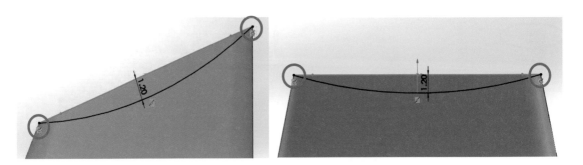

繪製後曲線如下。After drawing curves it looks following.

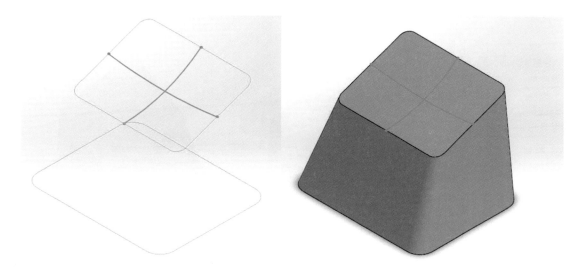

7. 利用以上繪製的兩條曲線，並運用邊界曲面功能產生一個曲面。Using boundary surfaces tool to create a surface.

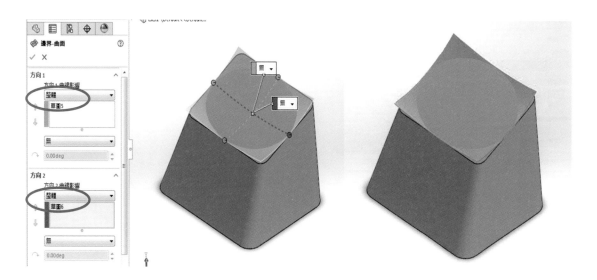

8. 運用置換曲面功能使鍵盤實體的上表面改變成曲面。Using Replace Face tool change body face to surface shape.

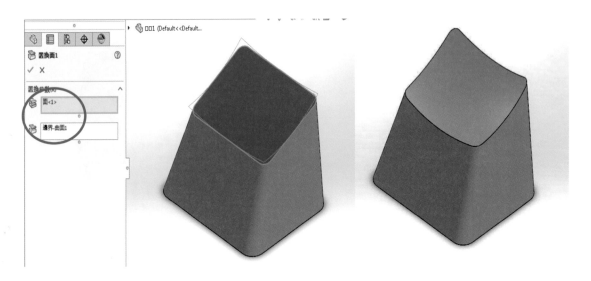

9. 產生 0.5mm 的圓角。Using fillet tool to make it smooth with 0.5mm.

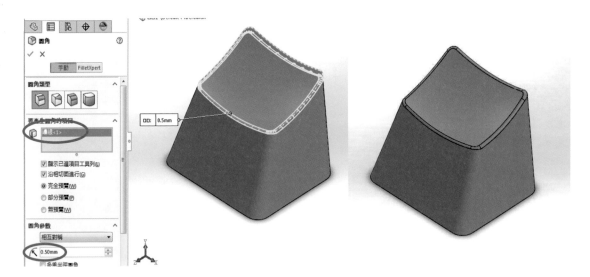

10. 產生 1mm 的薄殼。Create shell.

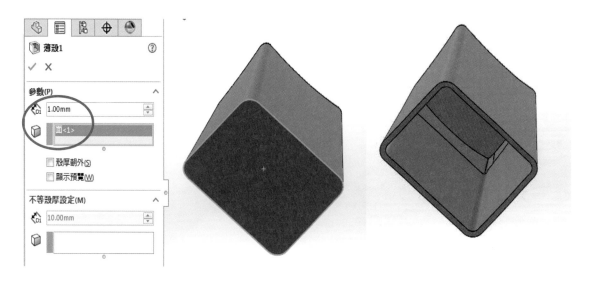

以下步驟為刻製文字。

11. 在上基準面利用草圖功能寫上所需文字，字體可以隨意變化。On the top plane, write the letter in sketch.

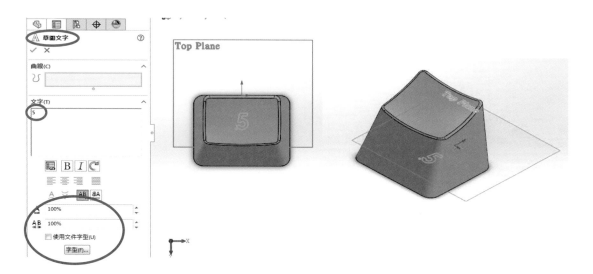

12. 運用敷貼功能將上基準面上的文字投影到鍵盤按鈕的表面，這裡字體高度定爲
0.25mm。Using wrap tool.

8.4 　螺旋把手造型 Spiral Handle Design

1. 在前基準面上繪製如下曲線（七點自由曲線）以及中心線，然後以中心線為旋轉軸產生如下的旋轉曲面。Sketch on front plane. Draw line using Spline. And using Revolved surface tool create a surface.

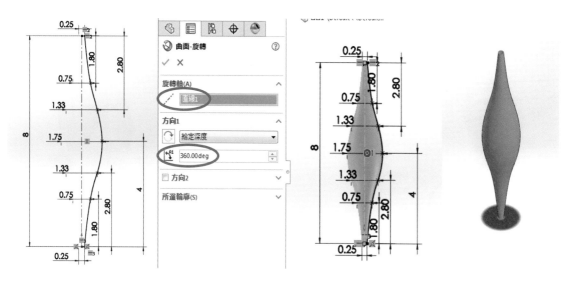

2. 在上基準面繪製如下草圖（一個 5mm 的圓）並產生一條螺旋線，參數如下。Sketch on top plane draw circle then using Helix and Spiral tool to make it spiral.

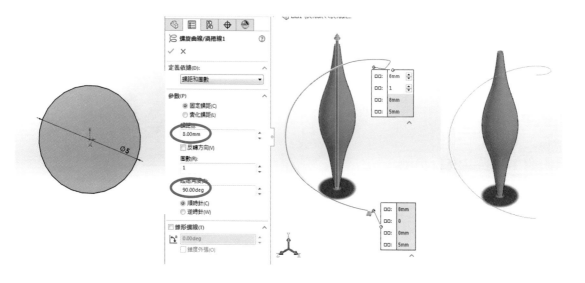

3. 在上基準面繪製另一個草圖（一條直線）並確認直線的一個端點與座標原點重合，另一個原點與以上的圓形重合，然後運用曲面－掃出功能產生螺旋曲面。Sketch on top plane draw line then using Swept boss/base tool.

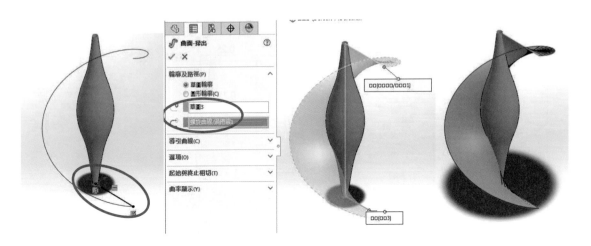

4. 運用相交曲線功能產生兩個曲面的相交曲線。Using intersection curve tool to get line.

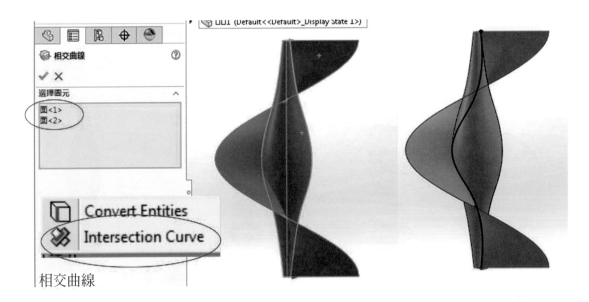

5. 在步驟1.曲線的端點產生一個平面,並在這個平面的中心繪製一個草圖(0.25mm 的圓,圓棒的粗細)。Create plane using spline then sketch on new plane draw circle.

6. 運用掃出功能產生曲線圓棒。Using swept boss/base tool to make it body.

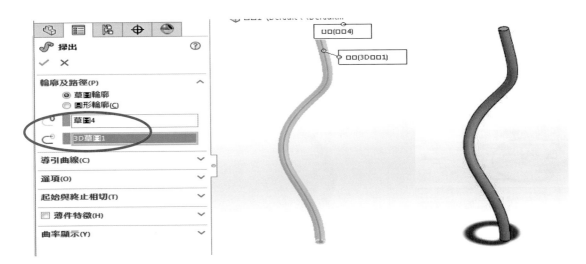

7. 在頂部平面上繪製草圖，並使用填料─伸長功能。Sketch on top plane draw circle and using extruded boss/base tool.

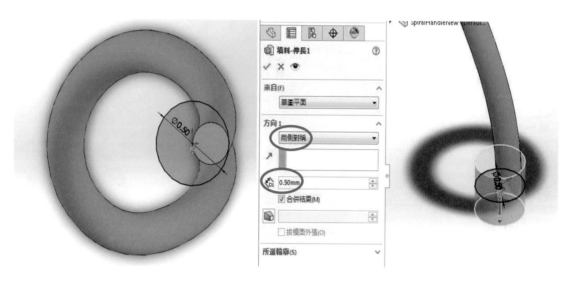

8. 運用圓形排列功能複製成以下的螺旋造型。Using circular pattern tools to duplicate.

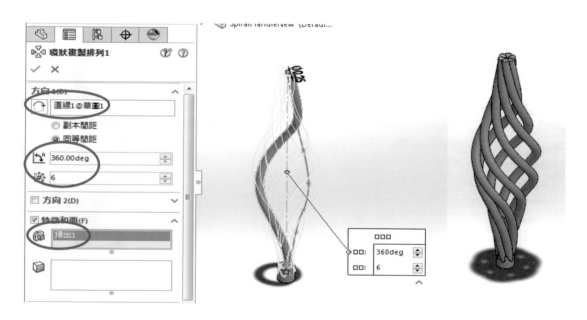

有了以上基本螺旋造型，讀者可以任意發揮其他外部的連接與功能，以下僅顯示一種，可以用作抽屜或閘門的手把。With the basic spiral solid model, the reader can make a lot of relative products such as handle of a drawer.

9. 創建一個新的基準面然後畫圓。Create new plane then draw circle.

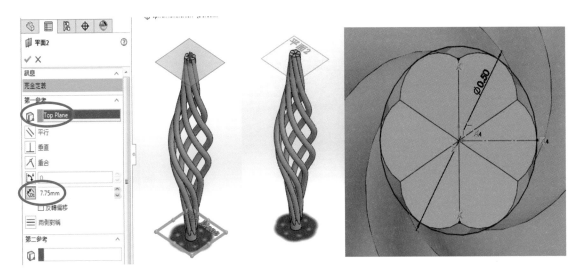

10. 使用填料—伸長功能。Using extruded boss/base tool.

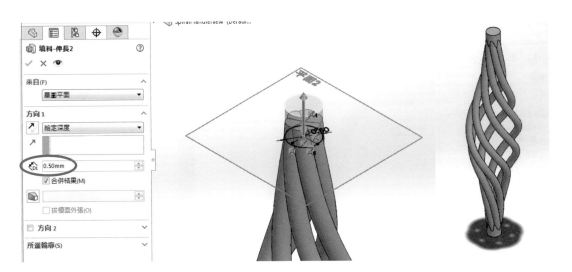

11. 在表面上繪製草圖並轉換成邊緣。並在前基準面畫一條線。Sketch on the face and convert the edge. Sketch on front plane, and draw line.

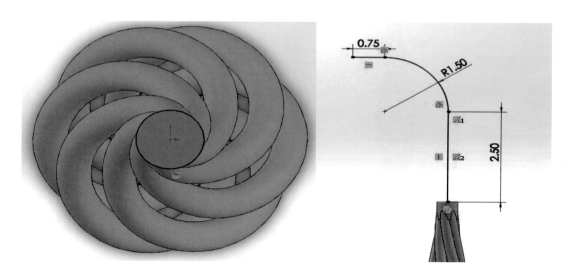

12. 使用掃出功能讓它堅固。Using swept boss/base tool to make it solid body.

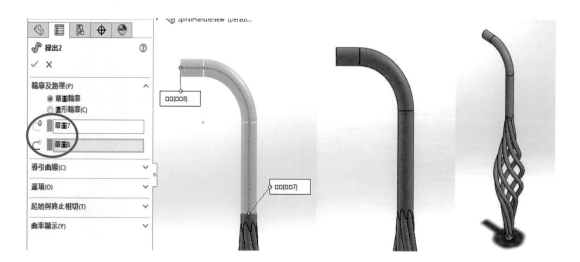

13. 創建一個新的基準面。Create new plane.

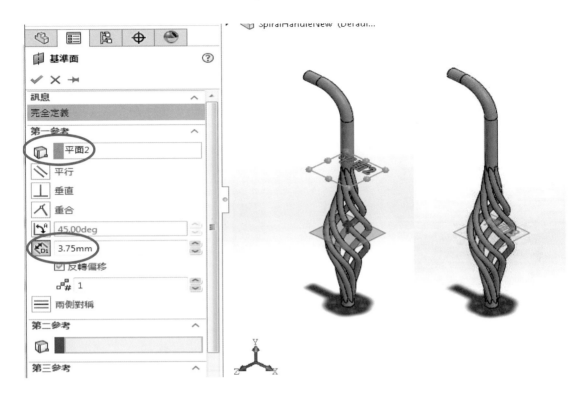

14. 使用鏡射功能。Using mirror tool.

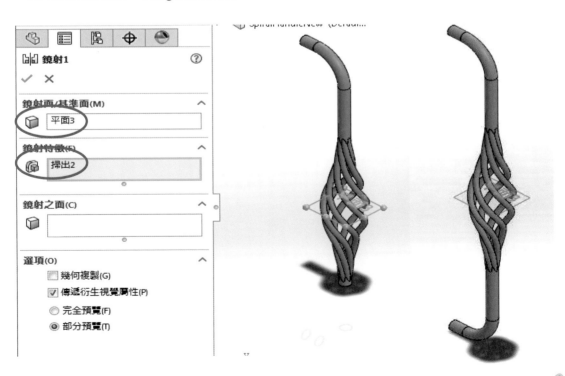

15. 使用圓角功能讓它光滑。Using fillet tool to make it smooth.

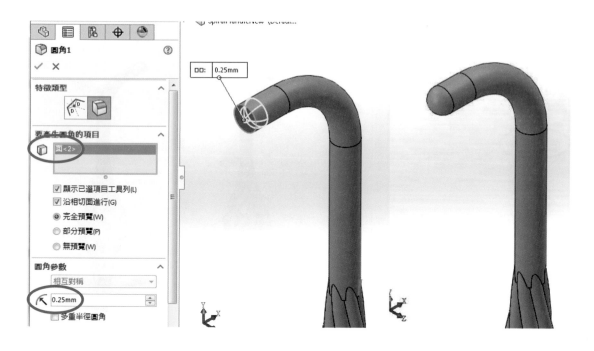

8.5　蜂窩環造型 Hexagon Mesh

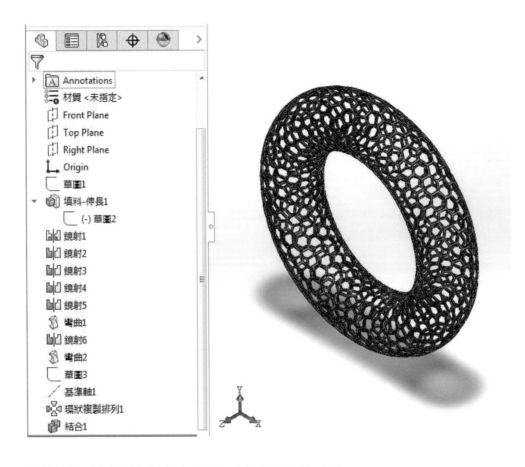

1. 在前基準面上繪製如圖的六角形，並且環狀複製三個。Sketch on front plane draw 6 sides polygon.

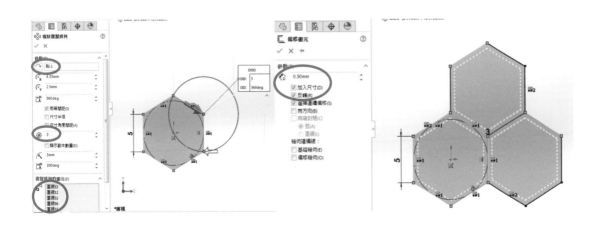

2. 利用剪切等工具將以上草圖改變成一個如下的 Y 形狀的封閉圖形。Draw line as shown below and using trim entities tools cut unnecessary sides.

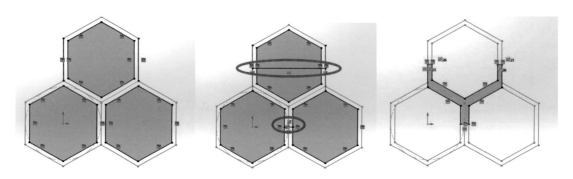

3. 將草圖伸長填料至 1mm 的實體。Using extruded boss/base tool to make it body.

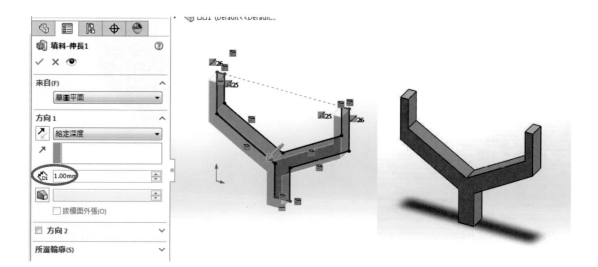

4. 數次運用鏡射工具產生以下的網格實體。Using mirror tool to duplicate the body as shown as below steps.

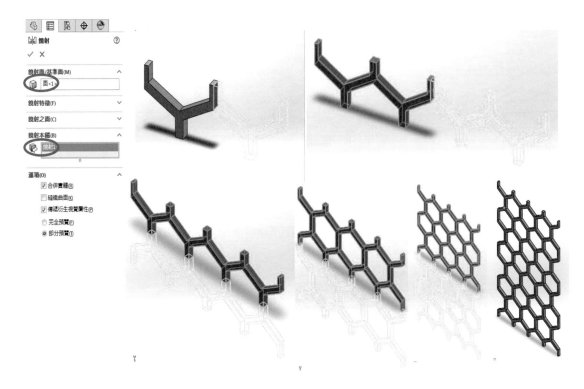

5. 運用彎曲工具使以上網格產生 180 度的弧形。Using flex tool to make it curve.

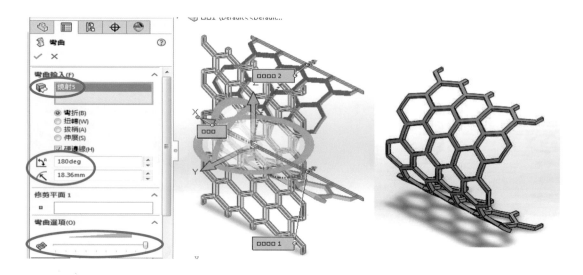

6. 運用鏡射工具將以上弧形網格產生成為圓柱形網格。Using mirror tool to duplicate the body.

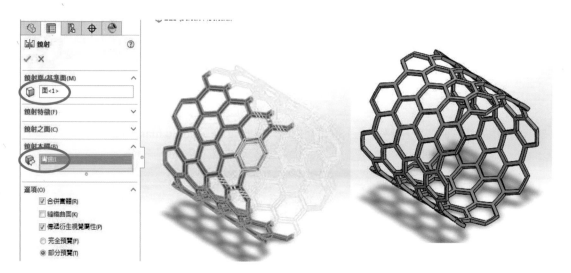

7. 再次運用彎曲工具將圓柱形網格以另一個方向彎曲成如下的形狀，具體參數為如下圖框選的 -30、90、90、90 度。Using flex tools make it around shape.

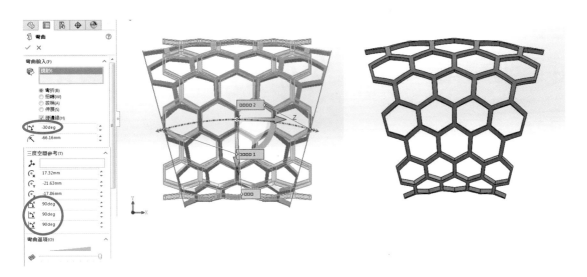

8. 在前視平面上繪製如圖的直線並找到交錯點，此交錯點位整個圓環的中心點。Sketch on front plane draw centerlines with infinite length as shown as below. Then put the point on the intersection and fix it.

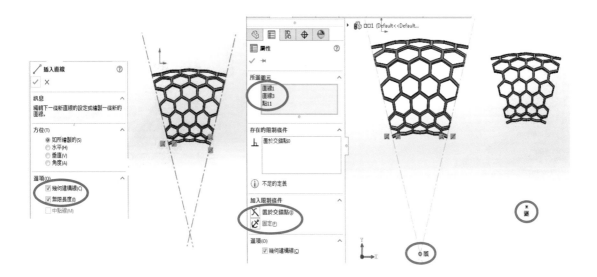

9. 利用以上的中心點產生一條基準軸。Create new axis using the point which created before.

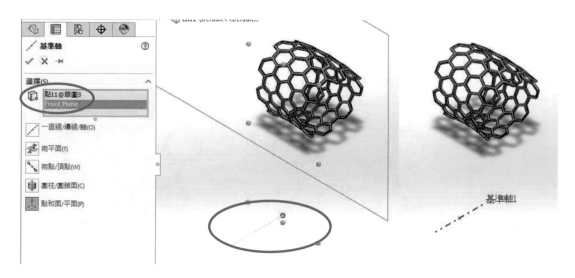

10. 這樣的話，再次運用環狀複製排列工具即可產生蜂窩環造型。Using circular pattern tool to make it full around body.

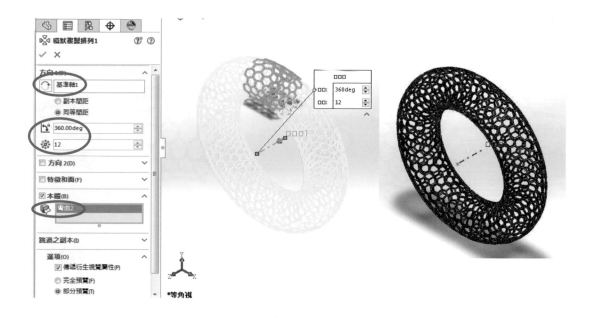

11. 最後，運用結合工具將所有實體組成一體。Using combine tool to make it one body.

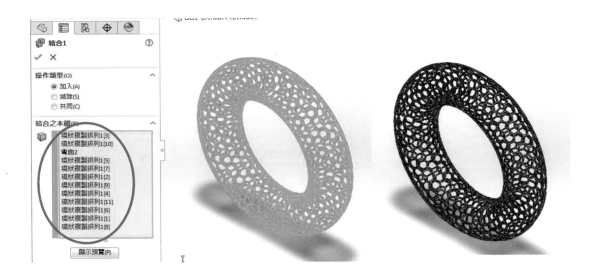

8.6 風扇罩殼造型 Fan Cover Design

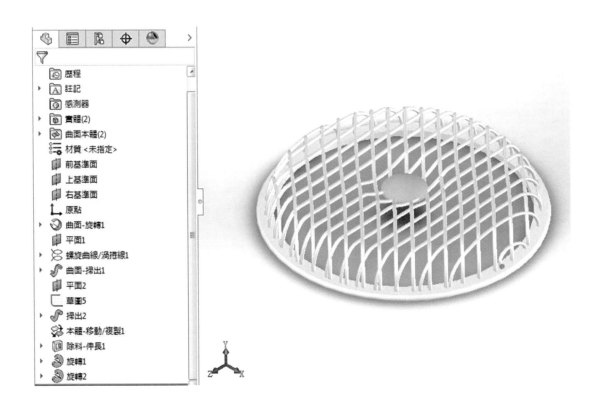

1. 在前基準面上繪製如下圖形，確保所有圖元之間的關係使之完全定義，然後運用旋轉工具，以草圖中心線為軸線產生曲面。Sketch on front plane. Then using Revolved surface tool to make surface.

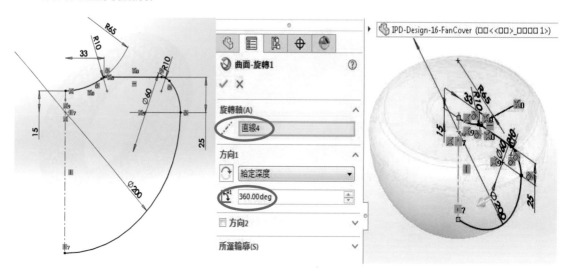

2. 產生一個新的基準面。Create new plane.

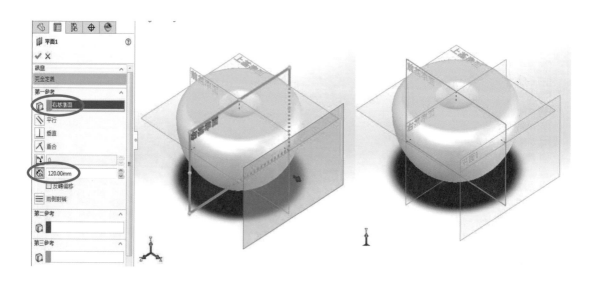

3. 在新的平面上繪製草圖（240mm 的圓）並產生一條螺旋曲線，參數如下。Sketch on new plane draw circle. Using Helix and spiral tools to make spiral curve.

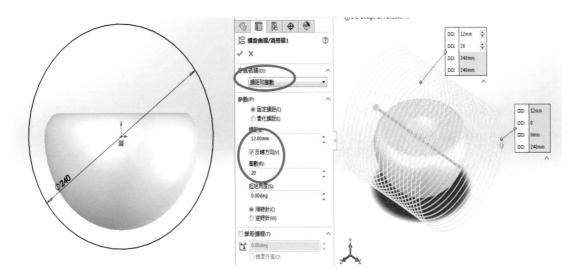

4. 在同樣的平面繪製另一格草圖（直線），直線的端點必須與以上草圖互相貫穿（使用的方法已經在前面章節敘述），然後運用掃出工具將直線與螺旋線產生一個如下的螺旋曲面。Sketch on the new plane draw line connect with curve relationship with pierce point. Then using swept surface tool to make it surface.

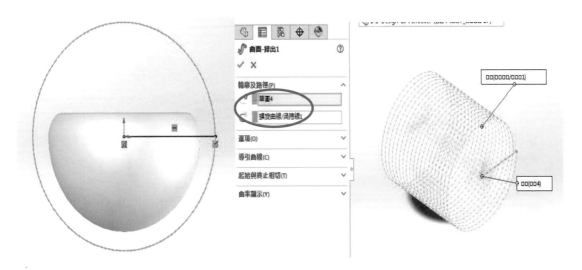

5. 產生一個新的基準面。Create a new plane.

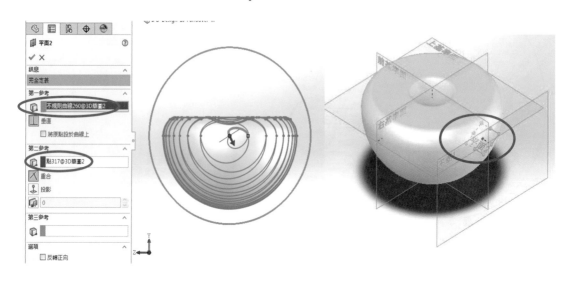

6. 在草圖工具列裡找出相交曲線工具，運用此相交曲線工具，產生以上旋轉曲面以及螺旋曲線的相交曲線如下。3D sketch on front plane.

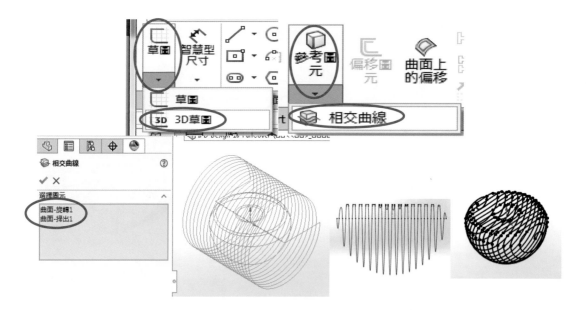

7. 在相交曲線端點建立一個平面（此時爲清晰起見可以將曲面隱藏），在此平面上繪製一個草圖（3mm 的圓）。Sketch on new plane draw circle.

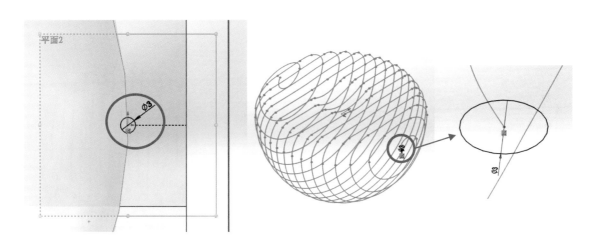

8. 運用實體掃出功能產生如下敷貼在旋轉曲面上的螺旋桿，作爲風扇罩殼的基本元素。

Using swept boss/base tool to make body using circle and 3d curve.

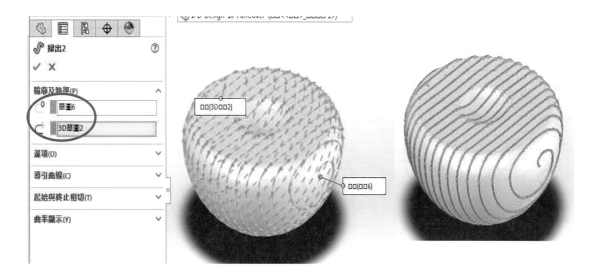

9. 運用移動／複製功能沿著 Y 軸複製 45 度的另一條螺旋杆。Using Move/Copy body tool to rotate body with copy.

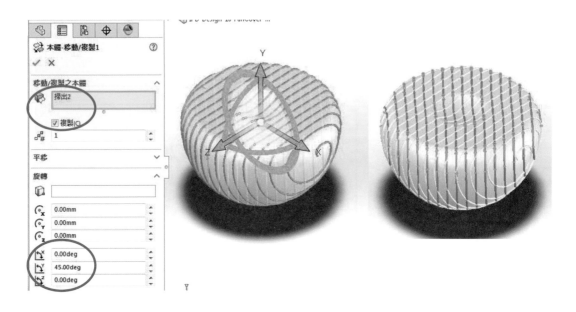

10. 以基準面為界，將無用的實體切除，形成了罩殼的主體。Using extruded cut tool to cut unnecessary part.

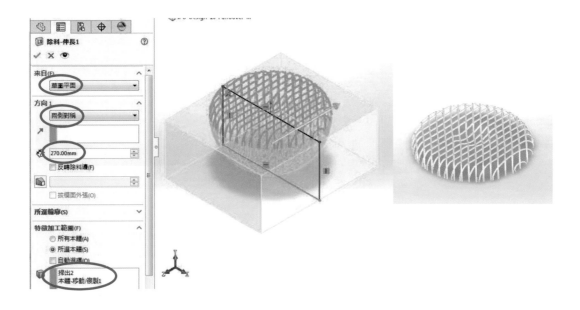

11. 繪製如下草圖。Sketch on front plane.

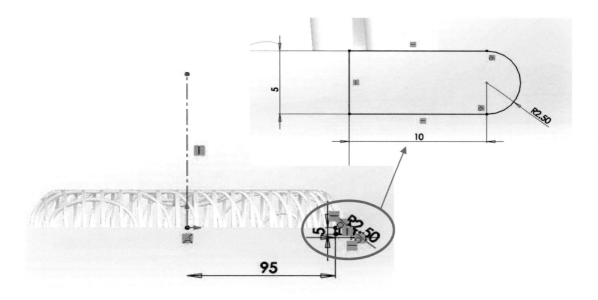

12. 運用旋轉工具產生罩殼的邊框。Using revolved Boss/Base tool to make it body.

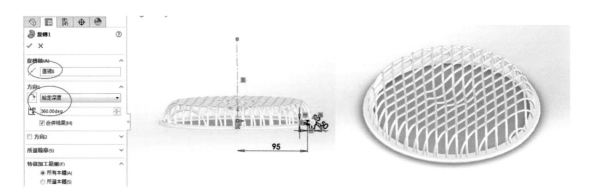

13. 再繪製一個草圖，並旋轉成如下的罩殼中心實體。Sketch on front plane. Then using Revolved Boss/Base tool to make it body.

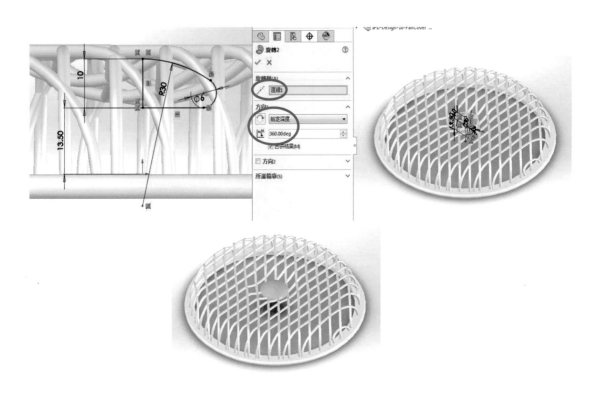

8.7 耳塞造型 Earphone Design

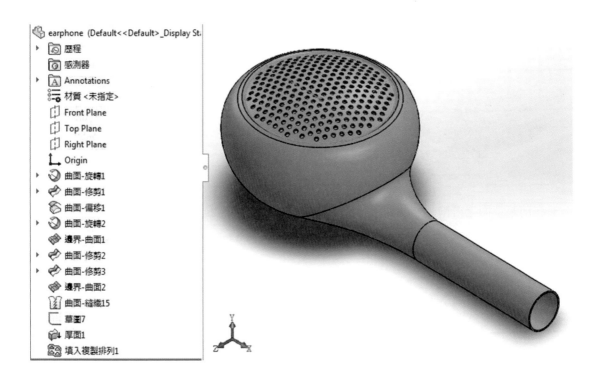

1. 在前基準面上繪製如下草圖，這裡請注意：此為三點自由曲線，並將曲線的尺寸及
角度修改成如圖所示，然後運用旋轉工具產生如下曲面。Sketch on front plane. Using
Revolved surface tool to make surface.

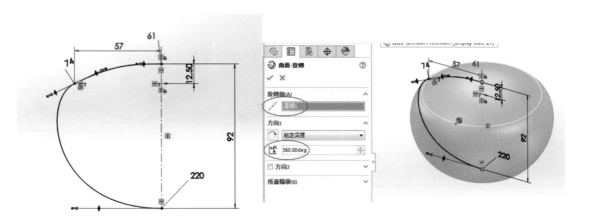

2. 在前基準面上繪製如下草圖，並以此切除旋轉曲面不必要的部分。Sketch on front plane. Using Trim surface tool to cut unnecessary side.

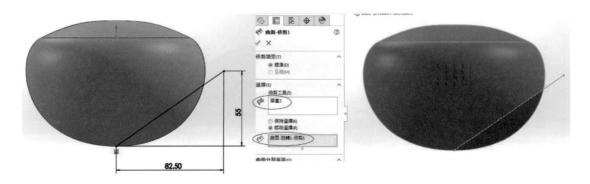

3. 在前基準面上再繪製一個如下草圖，並旋轉成一個圓柱面。Sketch on front plane. Using Revolved Surface tool to make surface.

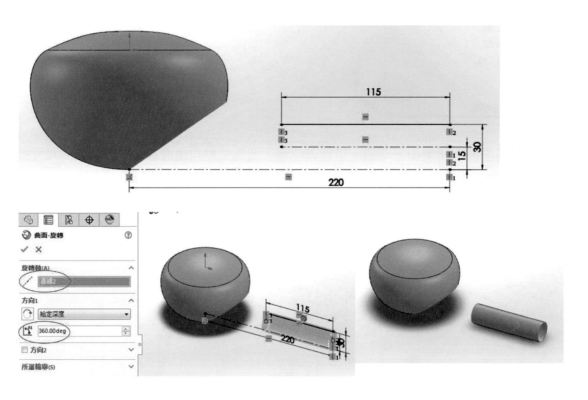

4. 運用邊界曲面工具產生如下的曲面。Using Boundary Surface tool to join 2 surfaces.

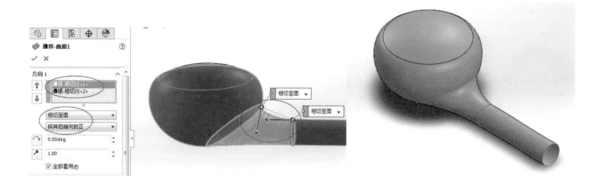

5. 運用偏移工具將上曲面往下產生偏移 2mm 的曲面。Using Offset surface tool to duplicate surface.

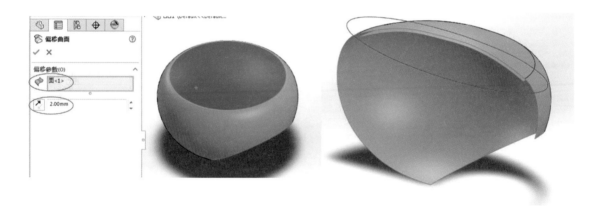

6. 在上基準面繪製一個草圖（108mm 的圓），然後利用此草圖將上曲面切除，露出剛剛利用偏移產生的曲面（圖示的紫色部分）。Sketch on top plane draw circle. Using Trim surface tool to cut unnecessary side.

7. 在上基準面再繪製一個草圖（104mm 的圓），然後利用此草圖將偏移曲面的外沿切除。

Sketch on top plane draw circle. Using Trim surface tool to cut unnecessary side.

8. 這樣就可以運用邊界曲面功能將間隙填滿。Using Boundary Surface tool to join 2 surfaces.

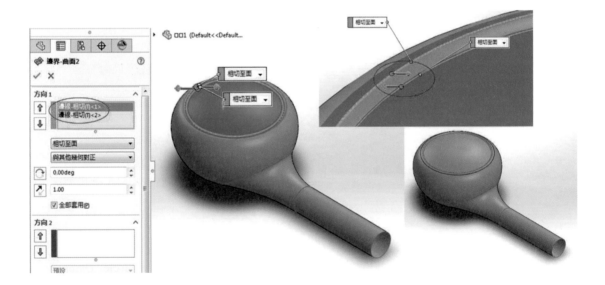

9. 運用縫織功能將所有曲面整合成一體。Using Knit surface tool to merge the all separate surfaces.

10. 運用厚面功能產生薄殼（此為 1mm）。Using Thicken tool to give thickness.

11. 在上基準面再繪製草圖，以為上曲面邊界為圓的邊界，並繪製一條中心線。Sketch on top plane convert face and draw centerline.

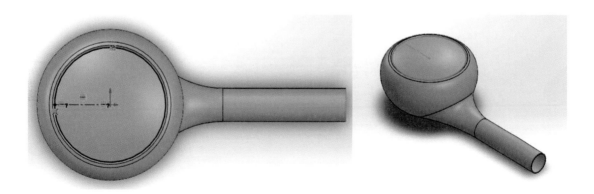

12. 運用填入複製排列工具產生耳塞如下的小孔。讀者可以嘗試各類排列選項，達到您滿意的排列方式。Using fill pattern tool to make a hole.

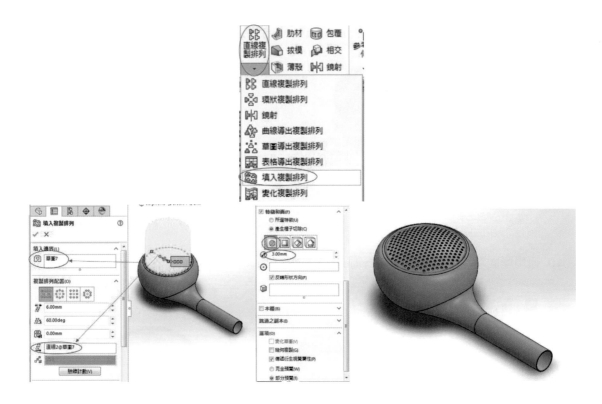

13. 特別注意的是，讀者還可以運用縮放（scale）功能將耳塞縮小到適合自己的尺寸。

8.8　腳踏車水壺支架設計 Bicycle Bottle Frame

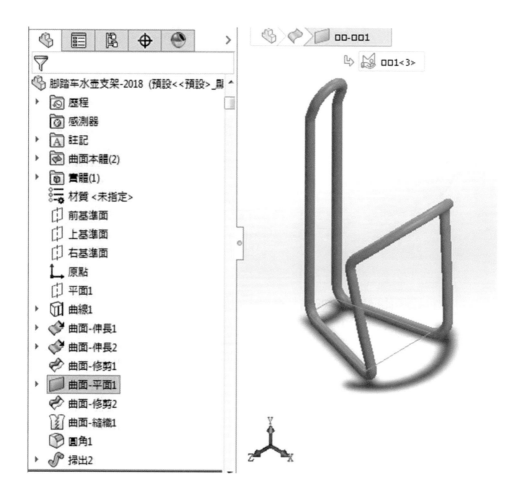

1. 在上基準面繪製如下草圖，運用曲面伸長功能產生一個圓柱曲面。Sketch on top plane. Use Extruded Surface tool with outside circle.

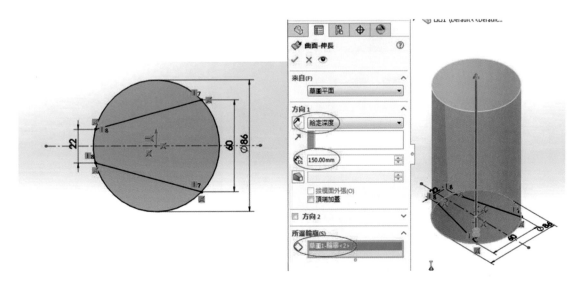

2. 產生平行於右基準面的草圖，並將其與以上草圖中的一個端點重合。Create new plane using right plane and point.

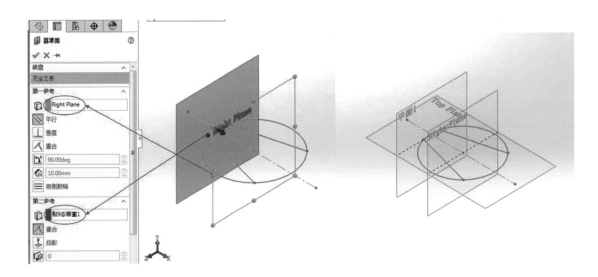

3. 然後在平面上分別繪製兩個草圖，請特別注意草圖尺寸及特徵的相關聯繫。Sketches on right plane and front plane.

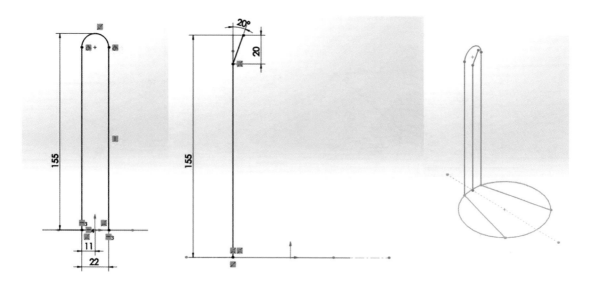

4. 運用投影曲線功能以及以上 2D 草圖，產生一個 3D 草圖。Using Projected Curve tools to create new curve.

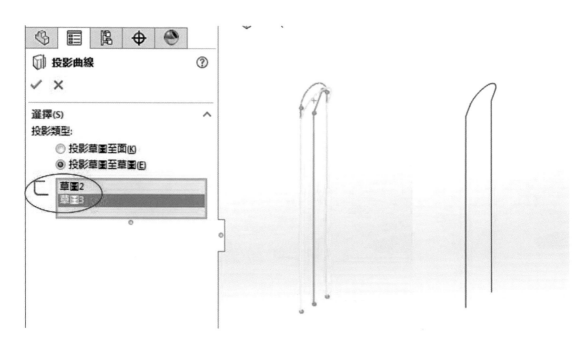

5. 在前基準面上繪製以下草圖。Sketch on front plane.

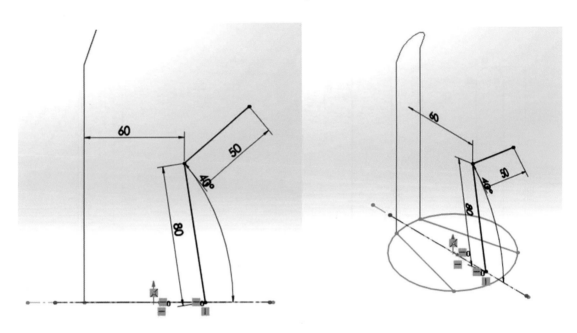

6. 運用伸長功能產生一個曲面（折面）。Using Extruded Surface tool to create surface.

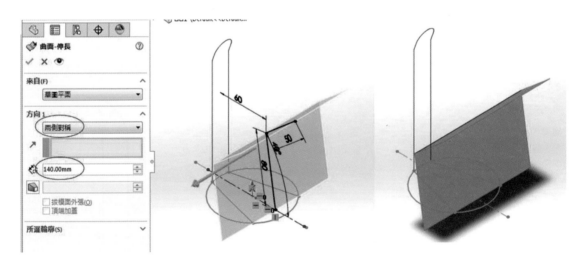

7. 運用曲面修剪功能，將剛剛產生的折面以圓柱面為剪刀裁剪。Using Trim surface tool to remove unnecessary part.

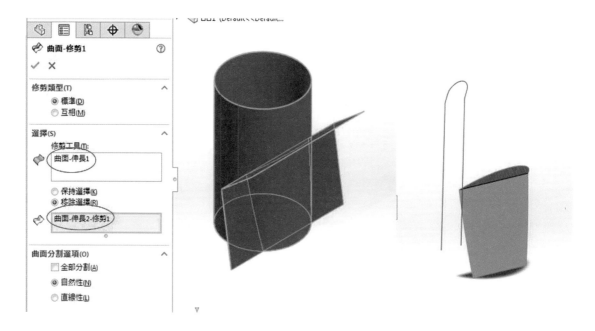

8. 在上基準面上繪製如下草圖，並產生一塊如下的平面。Sketch on top plane. Use Planar surface tool to create surface.

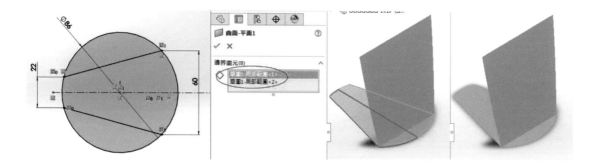

9. 運用修剪功能將剛剛產生的曲面進行修剪如下。Using Trim surface tool to remove unnecessary part.

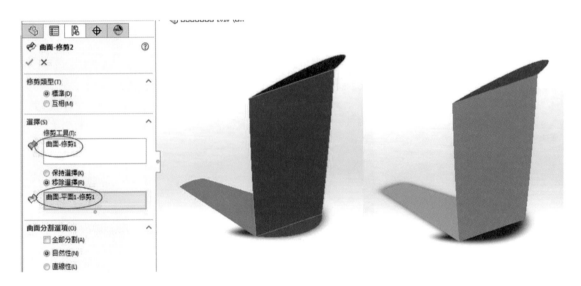

10. 再運用縫織功能將曲面拼成一體。Use Knit surface tool to combine surfaces.

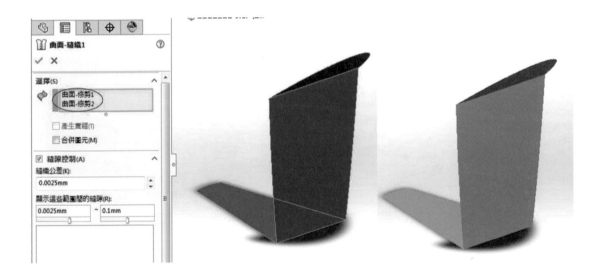

11. 產生 10mm 的圓角（曲面）。Use Fillet tools to make smooth.

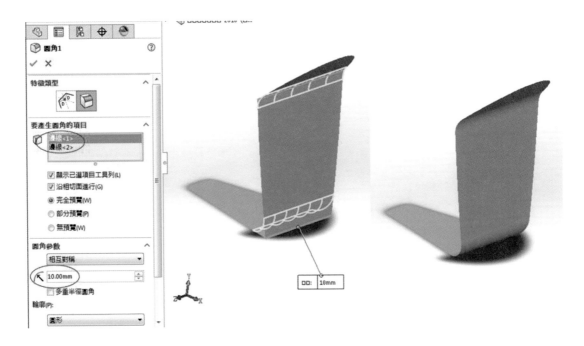

12. 繪製 3D 草圖，並利用轉化功能將以上曲面的邊界繪製成連續的 3D 曲線。Using 3D Sketch tool and then convert function to draw the continue 3D curves as below.

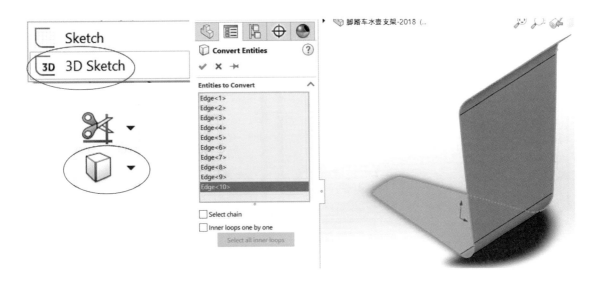

13. 在前基準面上繪製一個 6mm 的圓作為截面,這樣就完成了支架所有的基本元素。
Sketch on front plane draw circle. And convert tools to get curve.

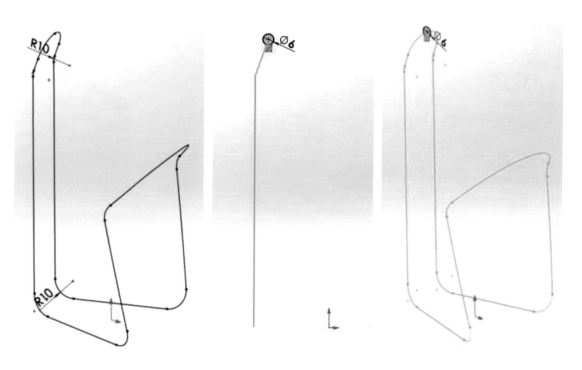

14. 運用掃出功能產生支架,並將不需要顯示的草圖及曲面隱藏,完成設計如下。Swept
Boss/Base tool to create body.

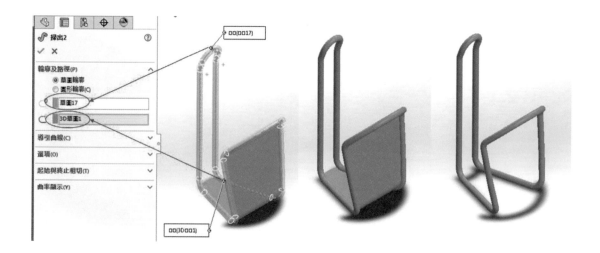

8.9 吊鉤造型設計 Design a Hook

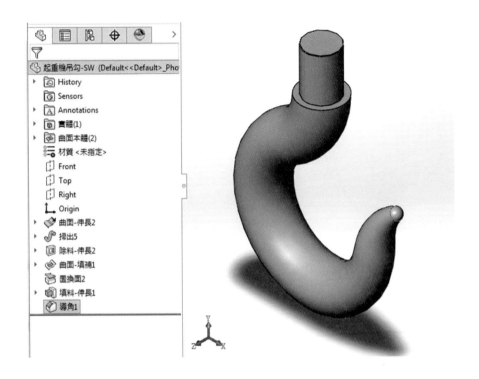

1. 在前基準面繪製如下草圖。Sketch on front plane as following below.

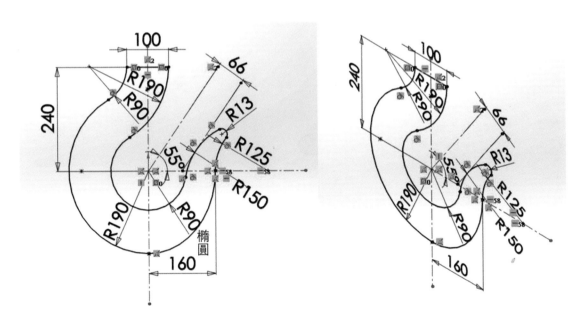

2. 在前基準面上繪製草圖（一條直線而已），可以運用草圖中的轉化功能以保證這條直線與以上草圖的關係。Sketch on front plane convert line.

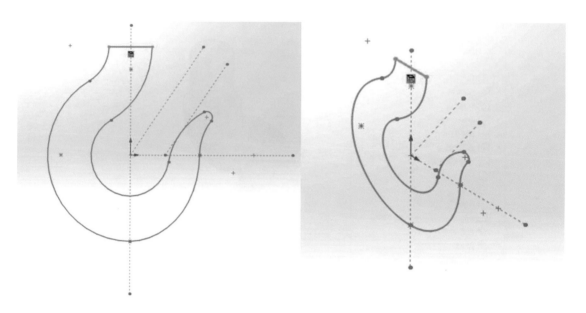

3. 運用伸長功能產生一個平面。Use Extruded surface tool to create surface.

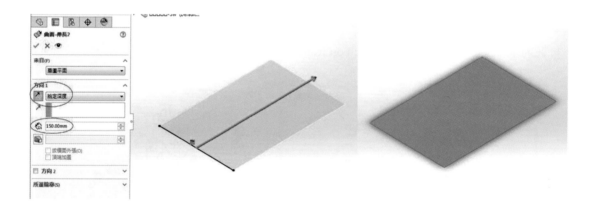

4. 在上基準面繪製一個草圖（一個圓而已），請特別注意圓的邊緣和第一個草圖的點相
交，然後開始運用掃掠（swept）功能產生實體。由於曲線是分段的，所以需要利用選
擇管理工具框（Selection manager）來進行。Sketch on top plane draw circle. Then Swept
Boss/Base tool to create body.

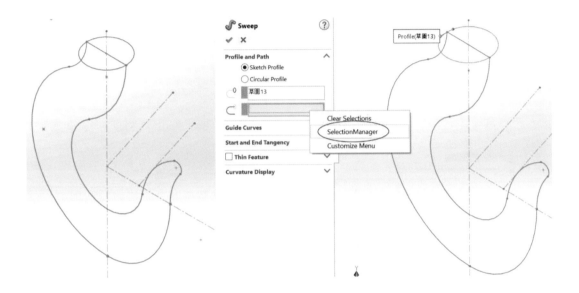

5. 引用了第一條曲線以後結果如下左圖，然後再定義引導曲線（Guide Curves），也是同
樣方式。The result of using first reference curve is as the left picture. Then you should define
the second reference curve as the picture on the right.

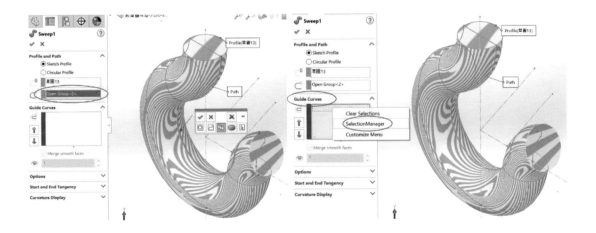

6. 選擇引導曲線，結果如下右圖。The result is like the following：

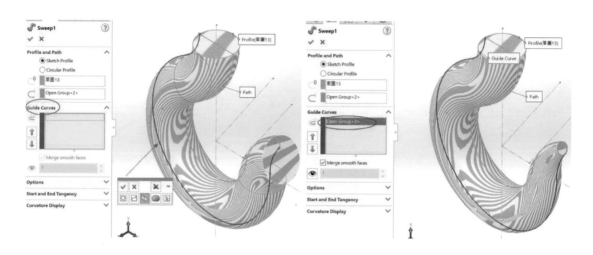

7. **以下步驟為修飾吊鉤的圓形端點。**首先在前基準面上繪製草圖（一條直線並與第一個草圖的兩個端點重合），並以此直線為剪刀切除多餘的實體。Sketch on front plane draw line. Using Extruded Cut tool to cut unnecessary side for more looks better.

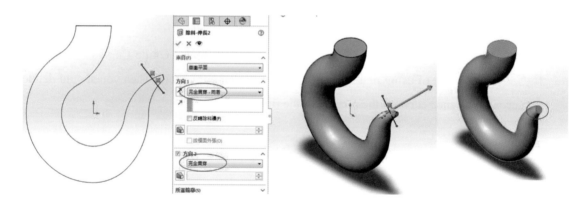

8. 在前基準面上繪製草圖（直線＋圓弧），運用填補工具產生半圓球形曲面。Sketch on front plane. Using Filled Surface tool to create surface.

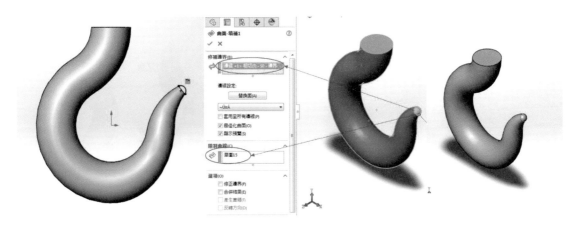

9. 然後運用置換面工具將端點部分填滿成實體。這樣吊鉤主體就完成了，置換面前後細節如下。Use Replace Face tool to replace surface to body.

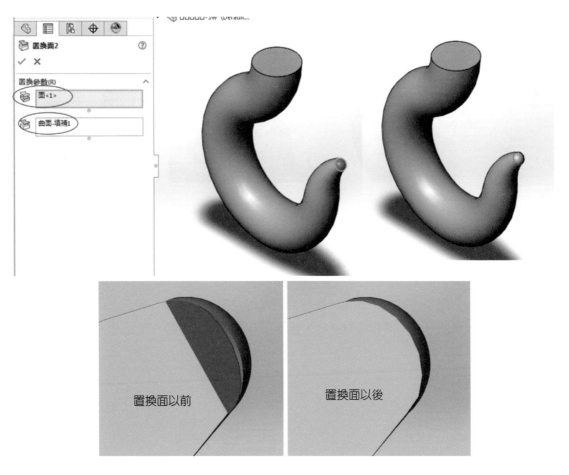

10. **以下步驟爲生成吊鉤的拉杆部分**。Sketch on top plane draw circle. Then using Extruded Boss/Base tool make body.

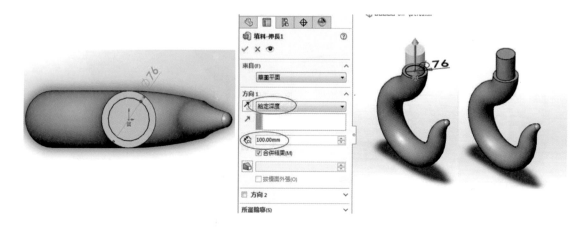

11. 運用導角功能 Use Chamfer tool.

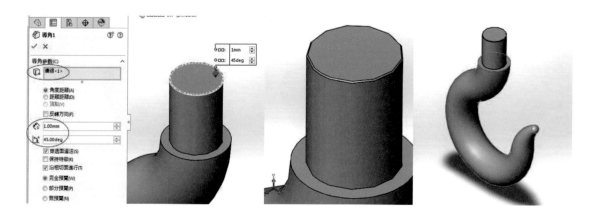

8.10   鑽石造型設計 Diamond

鑽石成型看似簡單，其實不簡單，由於關鍵點的位置，讀者也許會設計出各類造型，大家試一下吧。本章節只介紹鑽石造型部分，模具部分非常簡單，就不在這裡浪費篇幅了。

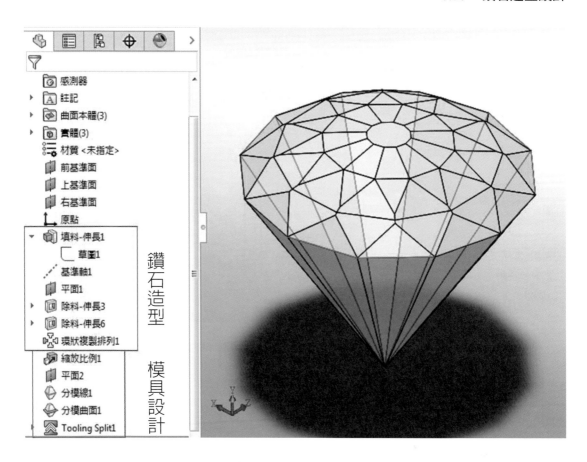

1. 在上基準面繪製草圖（一個 12 邊的多邊形），然後伸長為 22mm 的柱形。Sketch on top plane draw polygon. Then use Extruded Boss/Base tool to make body.

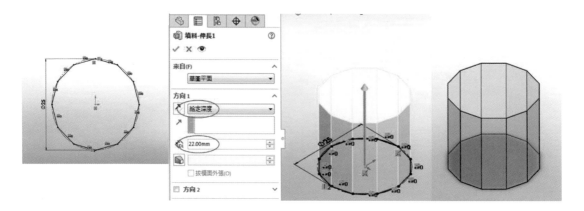

2. 利用兩個基準面產生基準線。Create axis using two planes.

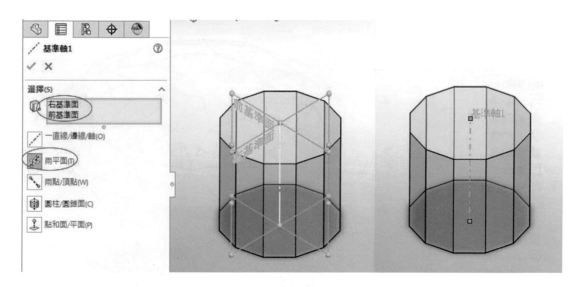

3. 在此平面上繪製如下草圖，然後運用伸長除料切除外部材料如下。Sketch on front plane. Then use Extruded cut tools to cut unnecessary side.

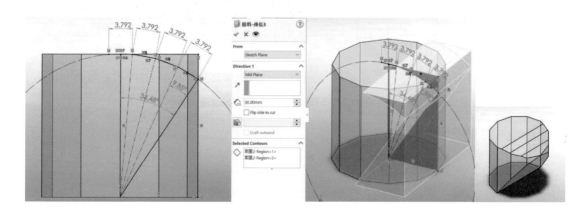

4. 產生一個基準面，注意以某一邊的中點為參考。Create new plane using axis which created before.

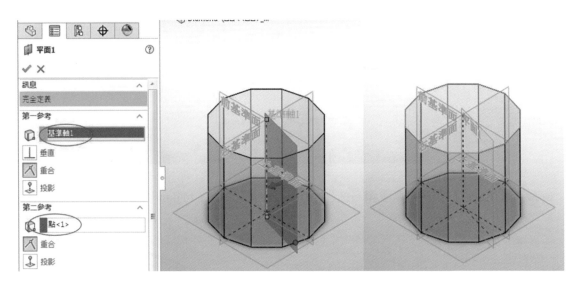

5. 在此平面上繪製如下草圖，然後伸長切除外部材料，請特別注意那些關鍵點的位置。Sketch on the plane. Then use Extruded cut tools to cut unnecessary side.

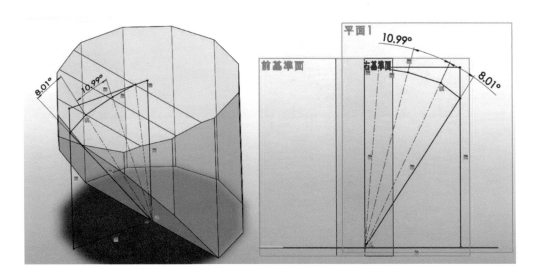

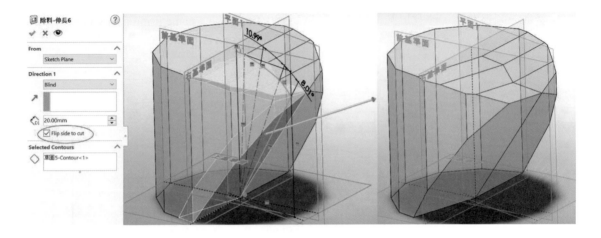

6. 最後運用環狀複製排列產生鑽石造型。Use Circular Pattern to duplicate body.

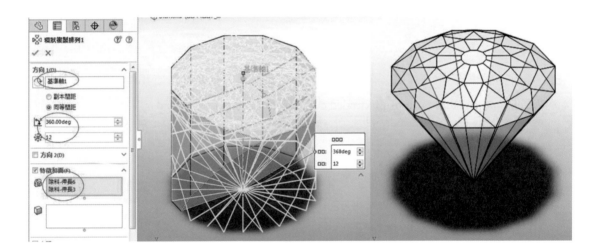

8.11　排氣管路聯結 Exhaust Manifold

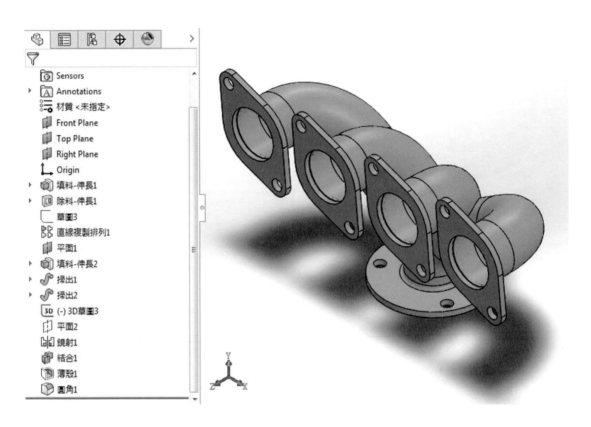

1. 在前基準面上繪製如下草圖，然後形成如下實體。Sketch on front plane. Use Extruded Boss/Base tool to make body.

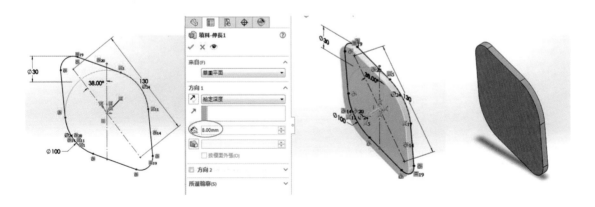

2. 還是在前基準面上繪製如下草圖，然後切出三個孔，實際上第一、二步可以併作一步成型。Sketch on front plane. Use Extruded Cut tool to cut unnecessary part.

3. 仍然在前基準面上，繪製一條中心線。Sketch on front plane. Draw centerline for next step.

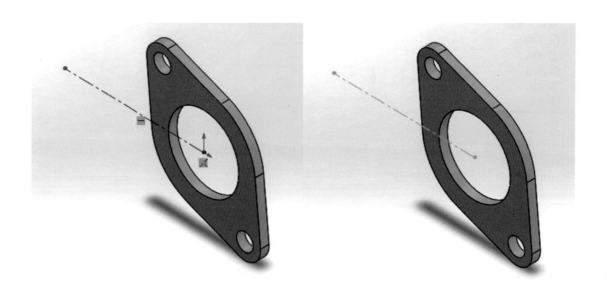

4. 用以上的中心線作爲直線複製排列的方向。Use Linear Pattern tool to duplicate body.

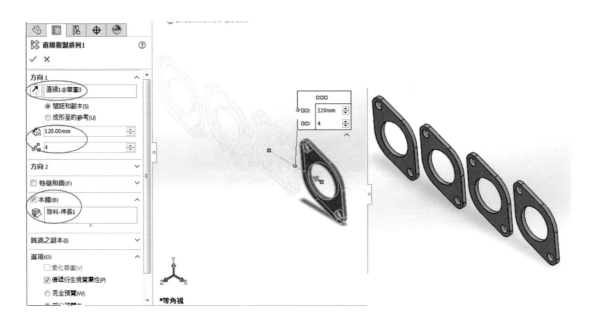

5. 產生一個新的基準面。Create new plane.

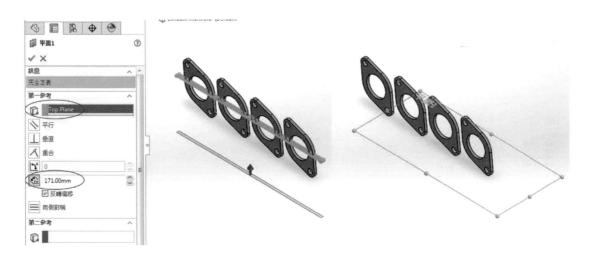

6. 在此基準面上繪製如下草圖，確保草圖內圖元之間的關係。Sketch on new plane.

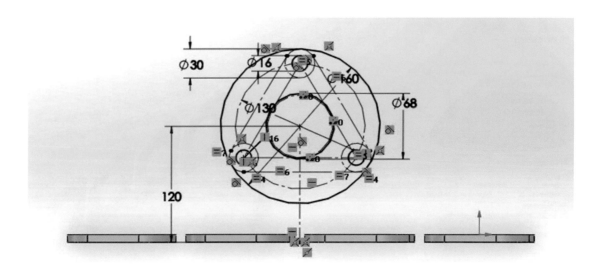

7. 以此草圖拉伸為一個法蘭圓盤。Use Extruded Boss/Base tool to make body.

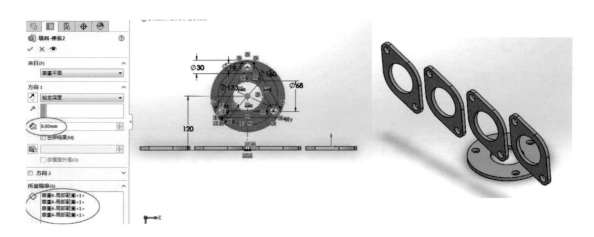

8. 先在法蘭盤上端平面繪製一個草圖（70mm 的圓），再繪製一個如下的 3D 草圖，最後
   運用掃掠成型爲一如下管路實體。Sketch on new plane draw circle. And use 3D sketch draw
   curve as shown below. Then use Swept Boss/Base tool to make body.

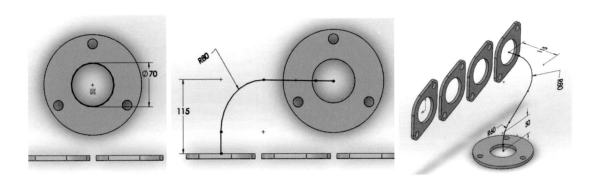

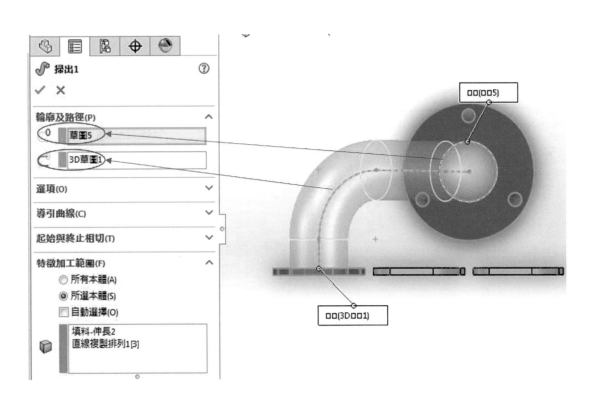

9. 以同樣的方法形成第二條管路。Same as previous step. Use 3D sketch draw curve and use Swept Boss/Base tool to make body.

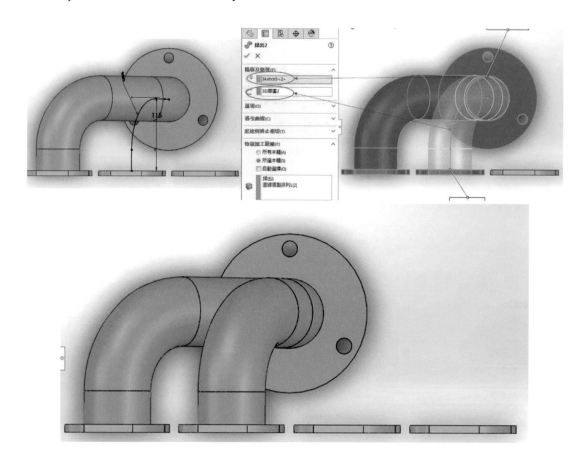

10. 以法蘭圓盤為中心繪製以下中心線。Use 3D sketch draw centerline without dimension for next step.

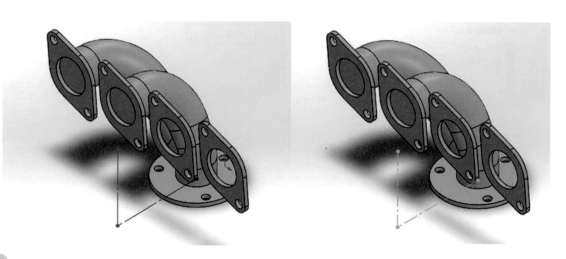

11. 利用此中心線建立平面 2。Create new plane using centerlines which created before.

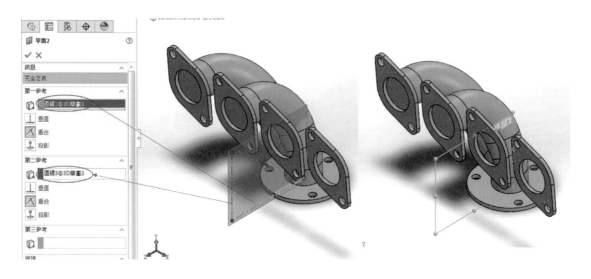

12. 以平面 2 為中心，將以上二管路鏡射。Use mirror tool to duplicate the body.

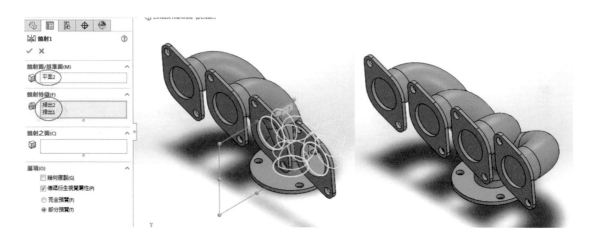

13. 運用結合功能將所有特徵結合爲一整體。Use combine tool to combine all separate bodies.

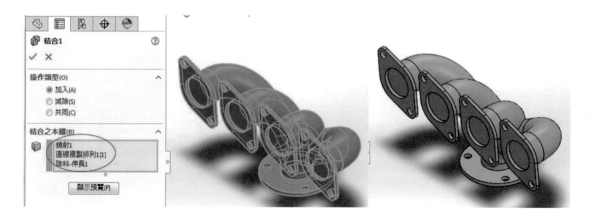

14. 運用薄殼功能將管路挖成中空。Use Shell tool.

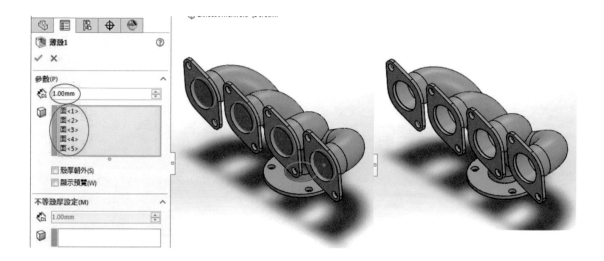

15. 在必要之處導圓角，完成設計。Use fillet tool make smooth.

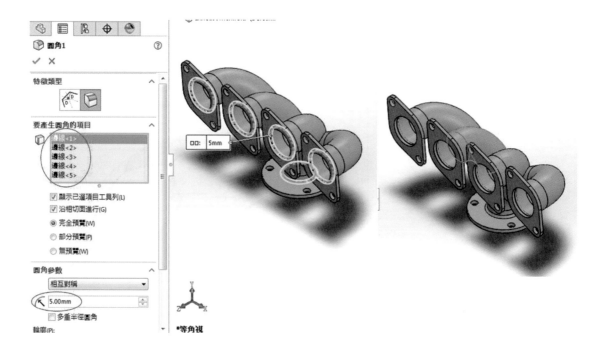

8.12　足球 Soccer Ball

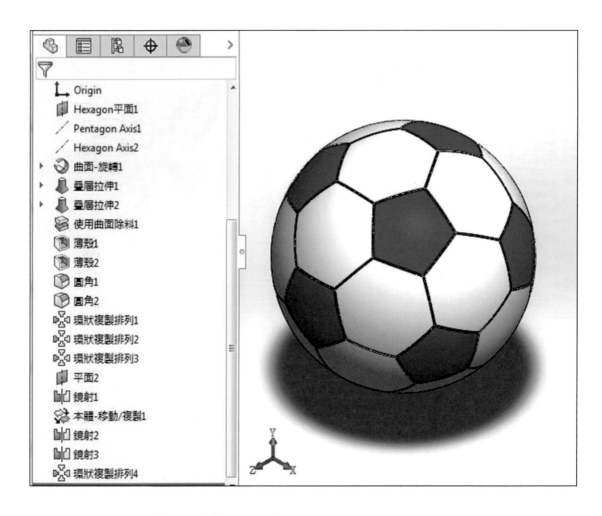

1. 在前基準面上繪製如下草圖（五邊形）。Sketch on front plane and draw 50mm Pentagon.

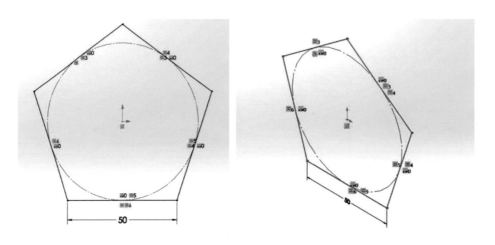

2. 建立一個 142.62 度的如下平面。Create new plane on any line of polygon with 142.62degree.

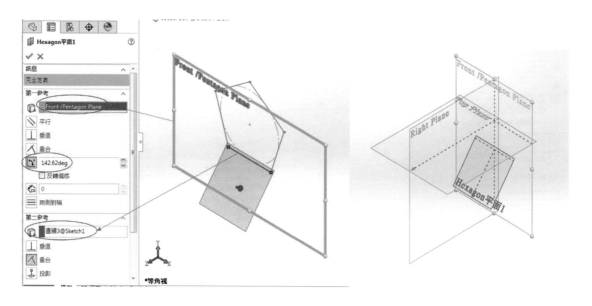

3. 在此新平面上繪製一個六邊形，確保五邊形和六邊形的一個邊共線。Sketch on new plane which created new one and draw a hexagon. Add relations with 2 lines collinear and equal.

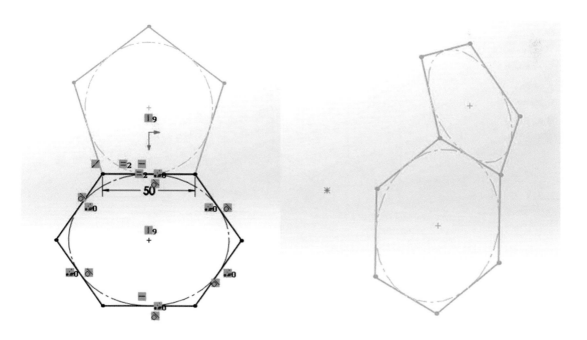

4. 以右基準面和上基準面產生相交軸線。Create new axis using Two Planes Right plane and top plane.

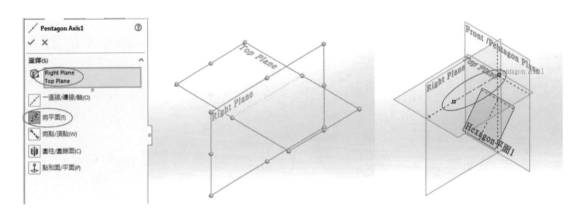

5. 產生一條輔助線，相關的關係如下圖所示。Create new axis using Point and Face/Plane tool.

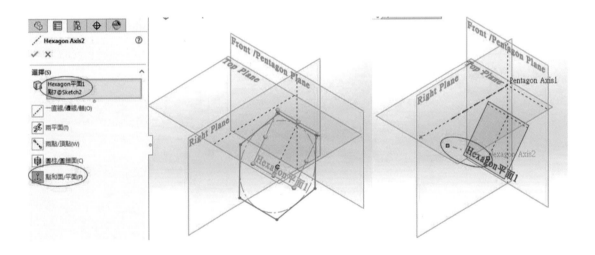

6. 運用 3D 草圖功能繪製兩個多邊形的法向焦點（僅一個點而已哦）。Choose the 3D sketch draw new point using 2 axis which created previous step and relationship should be coincident.

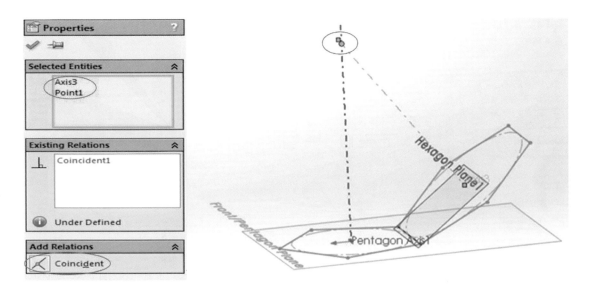

7. 先在右基準面上繪製一個草圖（220mm 的圓），然後以五邊形的軸心旋轉出一個球面。Sketch on right plane. Draw circle with 220mm diameter. Use Revolved surface tool to make surface. Make surface using pentagon axis.

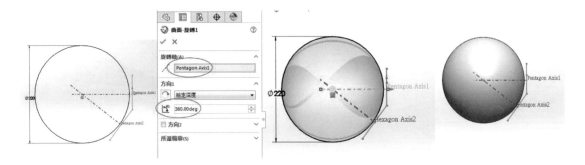

8. 運用疊層拉伸功能，以五邊形及中心點產生一個菱柱實體。Choose the Lofted Boss/Base tools and create solid body using point and pentagon polygon.

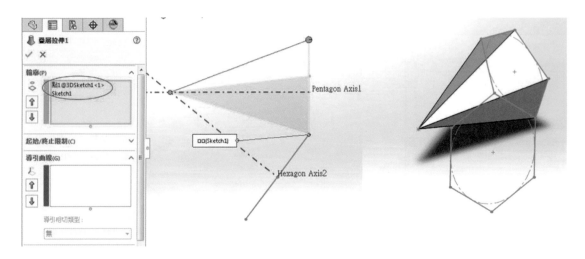

9. 以同樣方法成型六邊形的菱柱實體，請注意將「合併結果（R）」選項取消。Repeat step 8 but uncheck the Merge result tool.

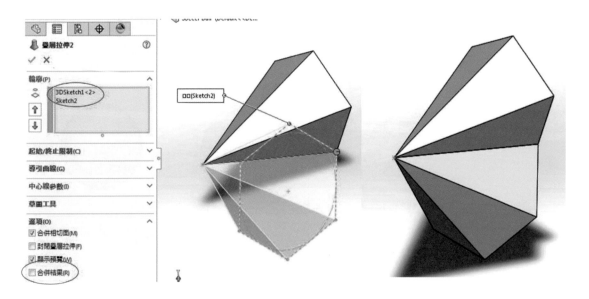

10. 運用曲面除料工具，用步驟 7 產生的曲面將實體內部挖空，如果曲面還在顯示則可以將其隱藏。Select Surface which created before and choose cut with surface tool.

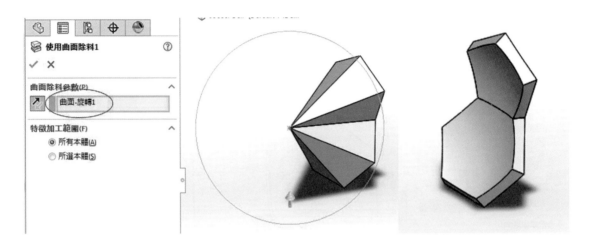

11. 運用薄殼功能將六角形部分型成 2mm 薄殼，請注意僅選擇六邊以及外表面，特別是選擇相交面時可以啟用透視功能及選項功能。Using Shell tool. Select all faces except surface. Thickness 2mm.

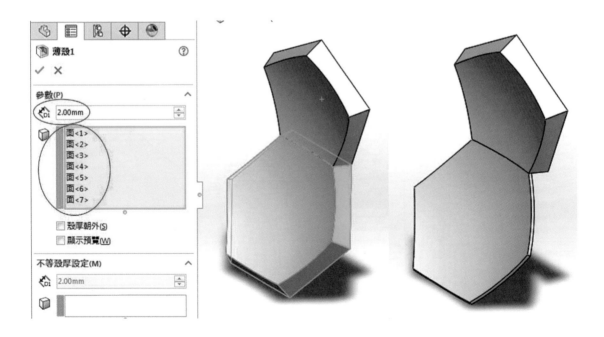

12. 以同樣方法處理五邊形薄殼實體。Repeat step 11.

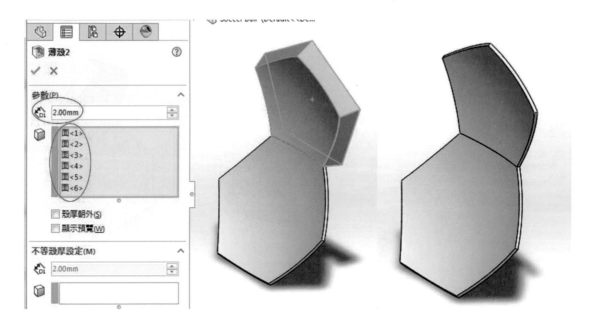

13. 將五邊形部分產生 0.75mm 的圓角。Make fillet with 0.75mm.

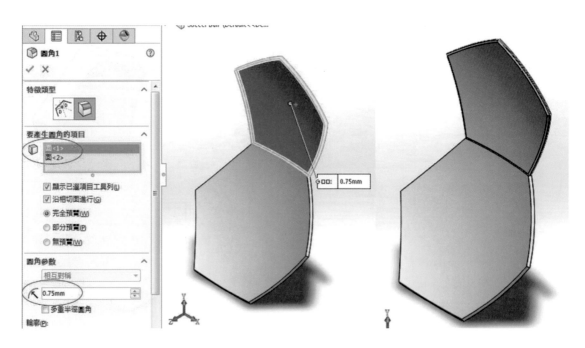

14. 將六邊形部分產生 0.75mm 的圓角。Repeat step 13.

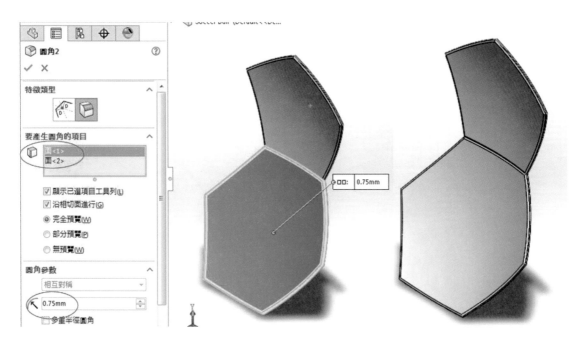

15. 運用環狀複製排列工具以及六邊形的法線，複製成 3 個五邊形，請注意選擇本體時要用滑鼠點選那個五邊形。Choose Circular Pattern tool.

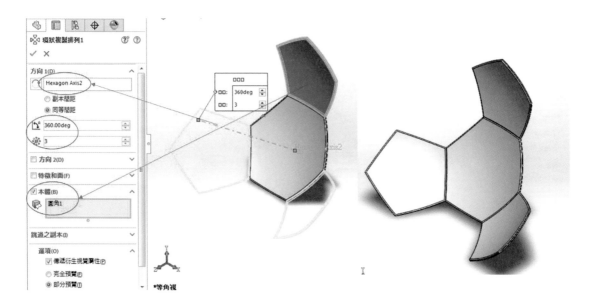

16. 運用環狀複製排列工具以及五邊形的法線，複製成 5 個六邊形，請注意選擇本體時要用滑鼠點選那個六邊形。Repeat step 15. Choose Pentagon axis.

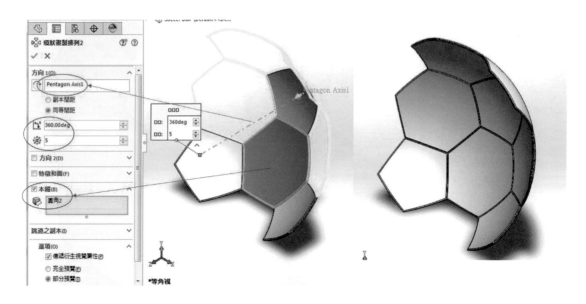

17. 再利用五邊形的法線環狀複製 5 個五邊形，請注意複製本體的選擇。Repeat step 15.

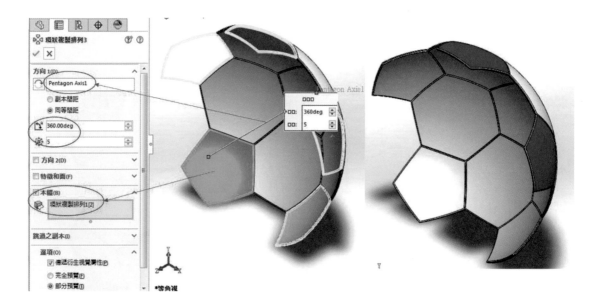

18. 利用球體中心點產生前基準面的平行平面 2。Create new plane using front plane and 3d point which created before.

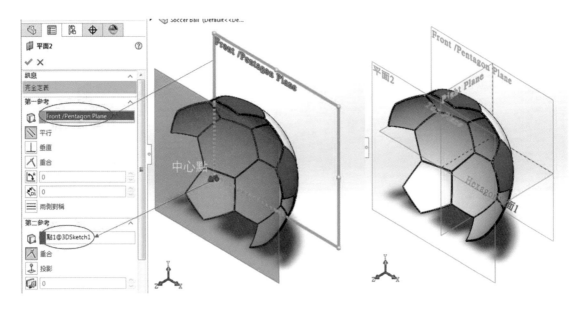

19. 運用鏡射功能，以平面 2 為基準，產生左邊的球殼部分，請注意選擇本體時要用滑鼠點選所有的多邊形。Choose Mirror tools.

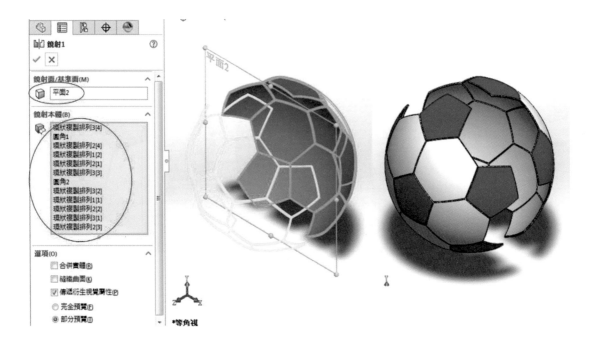

20. 運用本體移動／複製功能將左邊的球形部分沿 Z 軸旋轉 36 度，請注意此時可以方便地框選所需旋轉的多邊形。Choose Move/Copy Bodies tool to rotate the duplicated bodies.

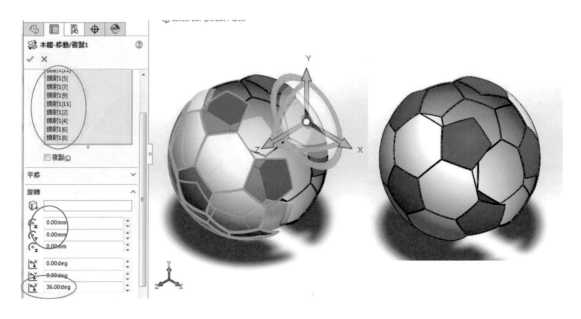

21. 鏡射（本體）一片六邊形。Choose mirror tool to duplicate the body.

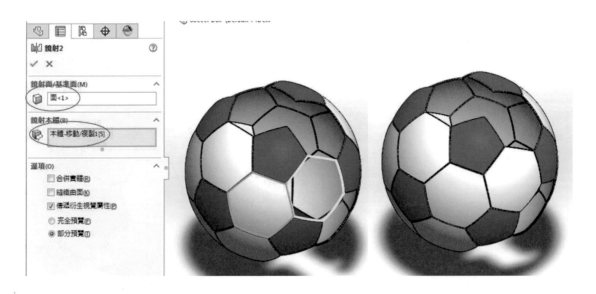

22. 再鏡射（本體）一片六邊形。Repeat step 21.

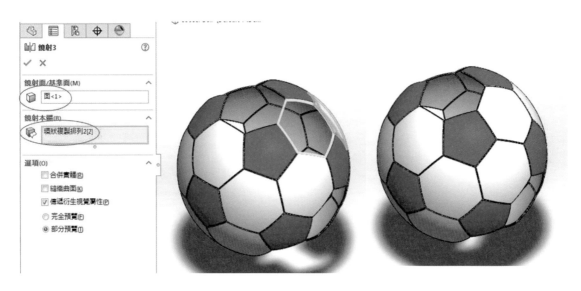

23. 以第一個五邊形的軸線為基準，將剛剛產生的兩個六邊形（本體）環狀複製排列，填補所有空白，完成足球的設計。Choose Circular Pattern repeat step 15.

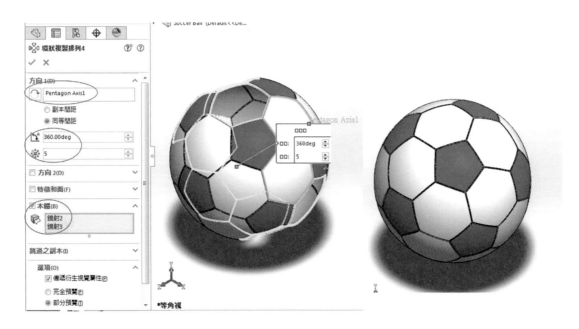

　　讀者也可以開啟並模仿光碟裡所附的平面形足球模型（8-12-02-RoughSoccer），步驟比較簡單。

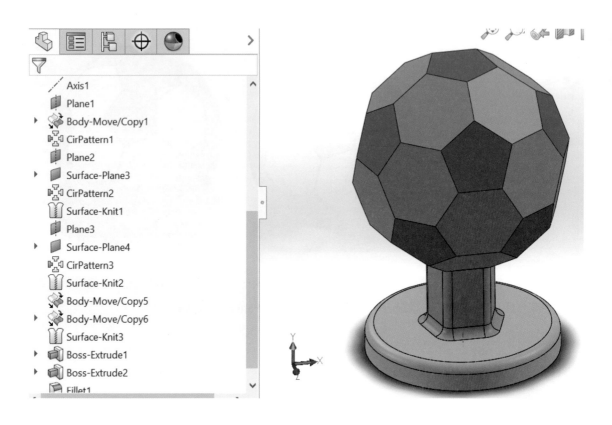

8.13 活動扳手設計及組裝 Adjustable spanner

8.13.1 扳手主體設計

1. 繪製草圖並拉深成厚度 10mm，注意兩邊對稱。sketch on front plane. Use Extruded Boss/ Base tool to make body.

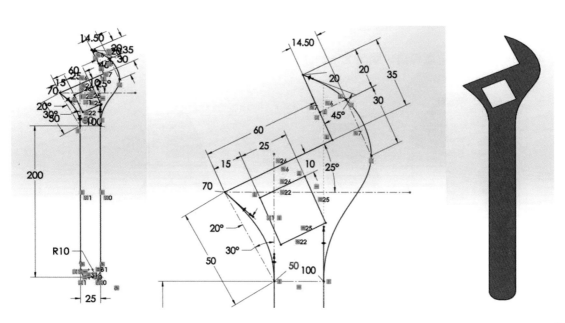

2. 在圖示之面繪製草圖並進行圖示的拉伸除料，方向 1 拉伸 5mm，方向 2 拉伸 35mm。

Sketch on surface of the body. Use Extruded Cut tool to make a hole on the body.

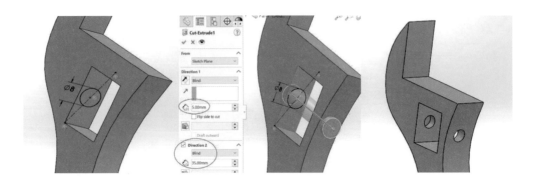

3. 在圖示之面繪製草圖並進行圖示的拉伸除料，方向 1 拉伸以及方向 2 均拉伸至下一面。

Sketch on surface of the body. Use Extruded Cut tool to cut the body.

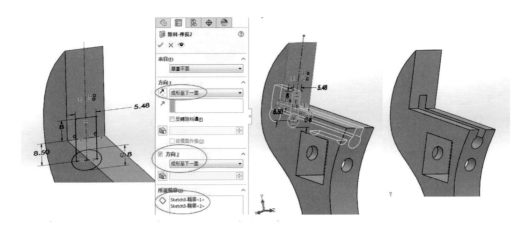

4. 導角：在圖示位置導角，距離 8mm，角度 60°。Use chamfer tool.

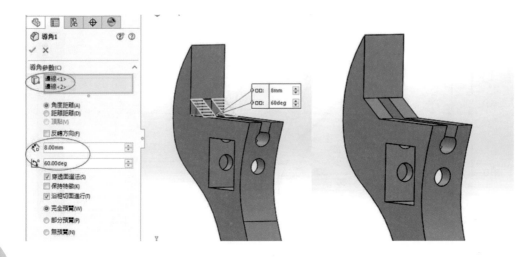

5. 鑽孔：在圖示之位置繪製草圖並進行圖示的拉伸除料。Sketch on surface of the body and draw circle. Use Extruded Cut tool to make a hole.

6. 周邊導圓角：選擇圖示的周邊，產生 2mm 的圓角。Use fillet tool.

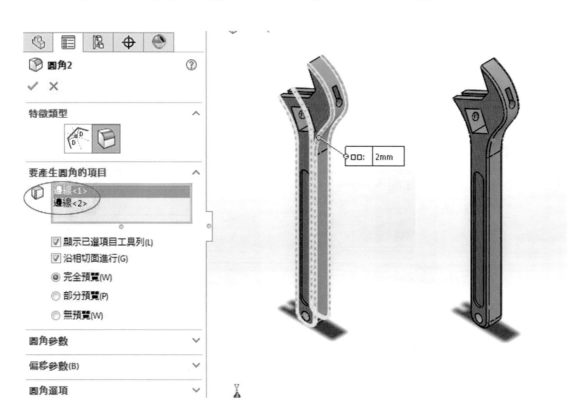

7. 產生凹陷：在圖示之位置繪製草圖並進行圖示的拉伸除料 2mm，並在底部產生 0.25mm 的圓角。sketch on the surface. Use Extruded cut tool.

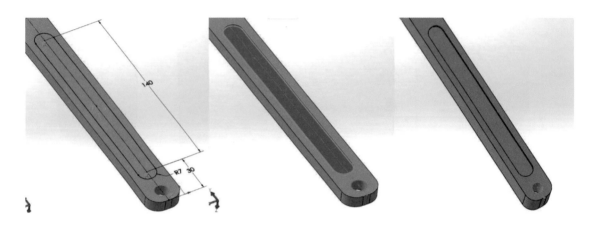

8. 鏡射：將以上的 2mm 除料及 0.25mm 導圓角特徵鏡射到另一面。Mirror the cut-off to the other side.

9. 雕塑字體：圖示位置繪製草圖，在草圖中沿著直線寫上文字，並拉伸特徵 1mm。Sketch on surface of the body write the text. Use Extruded Boss/Base tool.

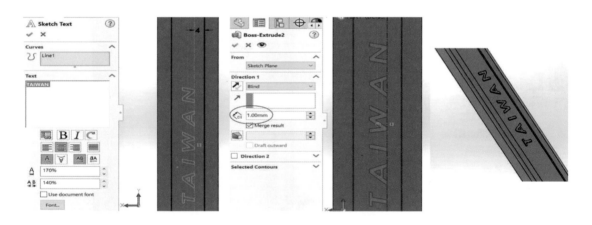

10. 背面可以雕塑不同的字體，完成成品如下圖。

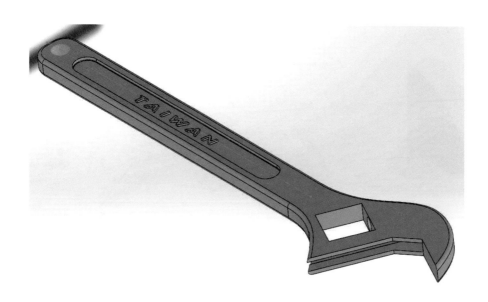

8.13.2 扳手滑塊設計（新零件）

1. 在前基準面上繪製草圖，雙向伸長成 10mm 實體。Sketch on front plane. Use Extruded Boss/Base tool to make body.

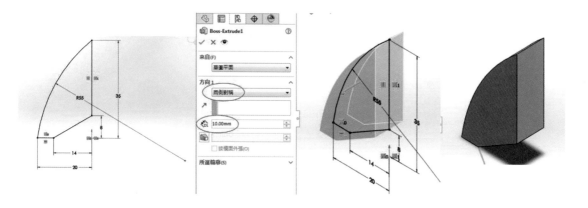

2. 同樣在前基準面上繪製草圖，雙向伸長成 5mm 實體。Sketch on front plane. Use Extruded Boss/Base tool to make body.

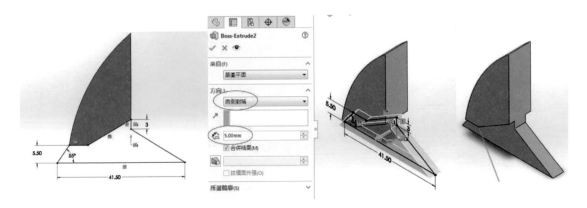

3. 在右基準面上繪製草圖（8mm的圓），伸長成圓柱，請注意方向1、方向2的不同長度。Sketch on right plane draw circle. Use Extruded Boss/Base tool to make body.

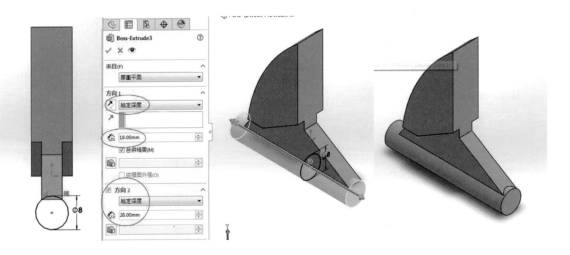

4. 在上基準面上繪製如下草圖，讀者可以利用線性複製排列工具，然後往下伸長除料，產生線性螺牙。Sketch on top plane. Use Extruded Cut tool.

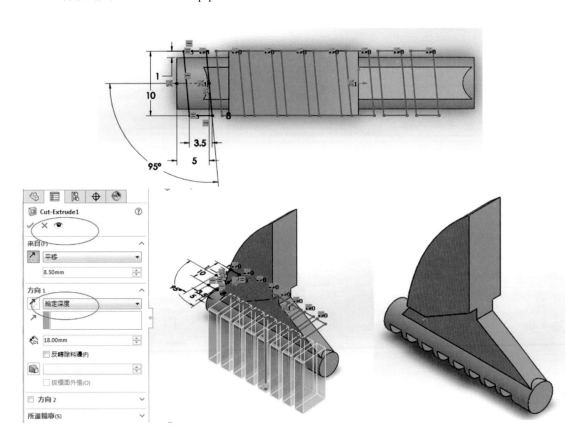

5. 在上基準面繪製草圖，再伸長除料切除不需要的部分。Sketch on top plane. Use Extruded Cut tool to cut unnecessary part.

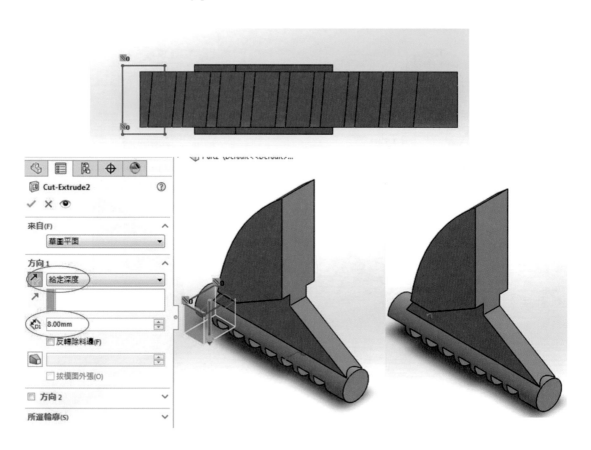

6. 在那一端產生 3mm 圓角。Use Fillet tool with 3mm.

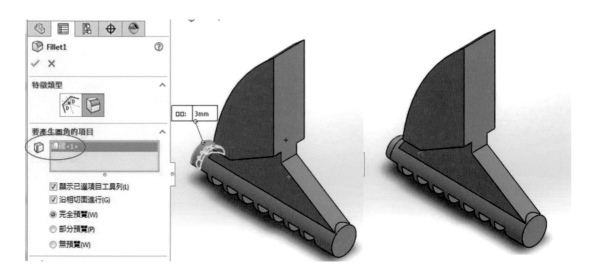

7. 在其他部分產生必要的圓角。Use Fillet tool.

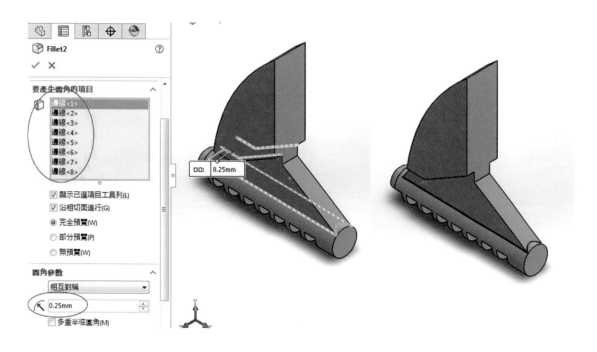

8. 在其他部分產生必要的圓角。Use Fillet tool.

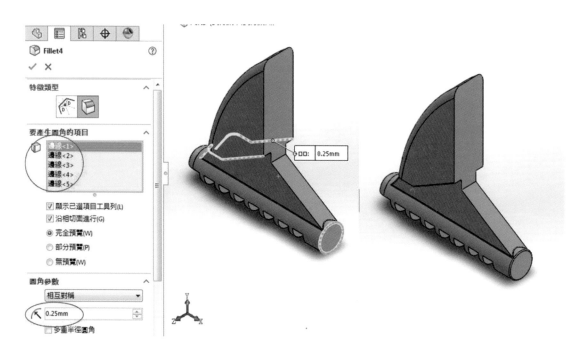

8.13.3 扳手螺母設計（新零件）

1. 在前基準面繪製草圖，並產生圓柱體。Sketch on front plane. Use Extruded Boss/Base tool.

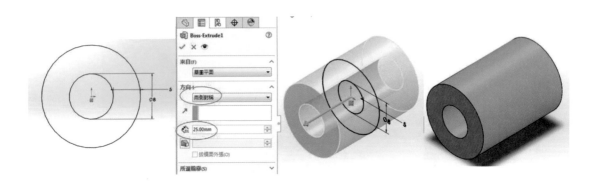

2. 在圓柱表面繪製草圖，並產生如下的螺旋線。Sketch on surface of the body draw circle. Use Helix/Spiral tool to make spring curve.

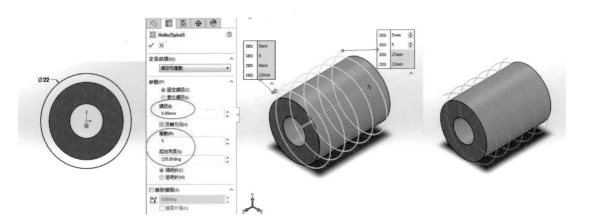

3. 在螺旋線的一個端點產生一個平面。Create a new plane.

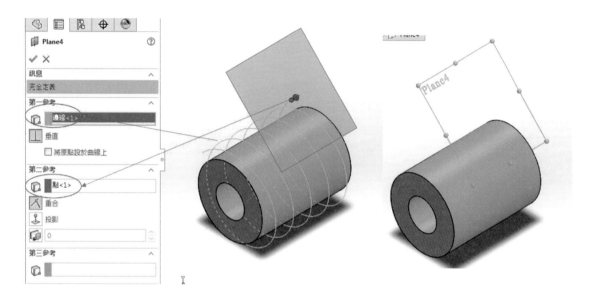

4. 在此新產生的平面上繪製如下草圖（3×3mm），並運用掃掠除料（Sweep）產生螺紋。
   Sketch on new plane. Use Swept boss/Base tool.

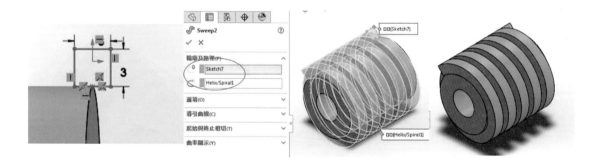

5. 在圓柱面上繪製草圖（一個足夠大的圓），並運用此草圖切除不必要的實體。Sketch on surface of the body draw circle. Use Extruded Cut tool to cut unnecessary part.

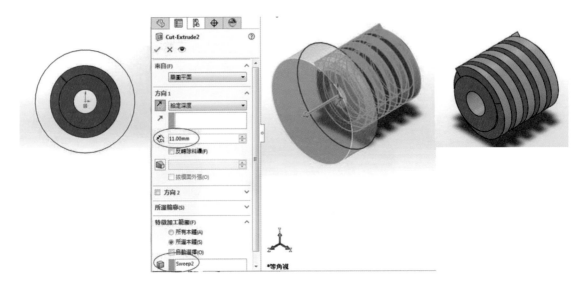

6. 運用鏡射功能除料另一邊的材料。Use mirror tool to cut another side.

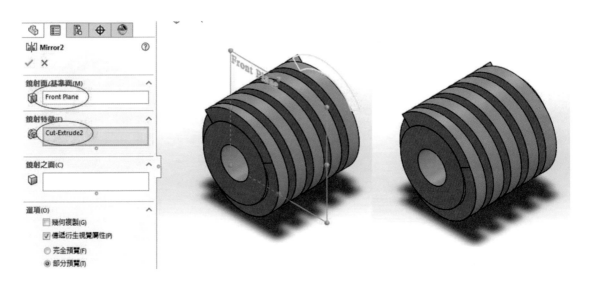

7. 導角 0.75mm。Use Chamfer tool.

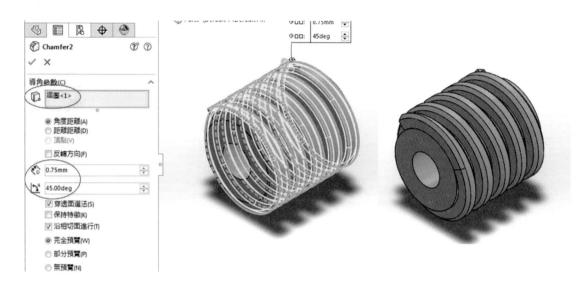

8.13.4 扳手螺母軸設計（新零件）

1. 在前基準面上繪製草圖（一個圓），並伸長成 35mm 的圓柱體。Sketch on front plane draw circle. Use Extruded Boss/Base tool to make a body.

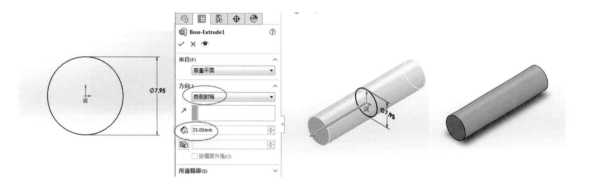

8.13.5 扳手組裝 Assembly

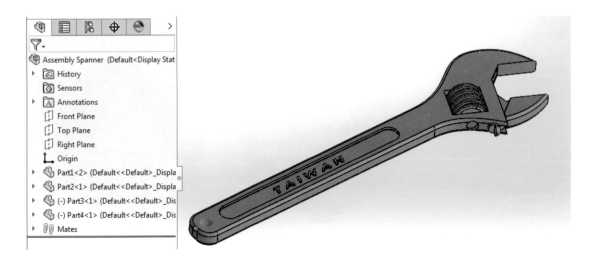

1. 開啟一個新的組合件。Create new file select assembly.

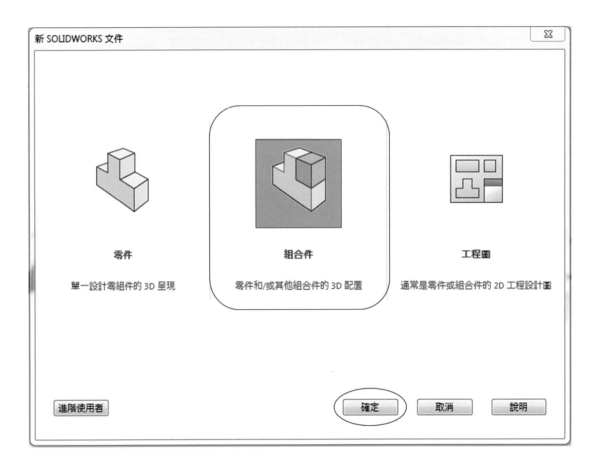

2. 插入所有的零件，請注意首先插入扳手主體（使之成為固定）。Insert the all parts.

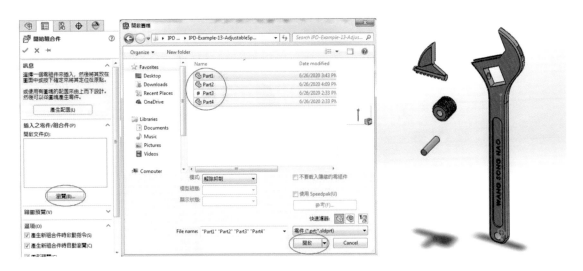

3. 將扳手螺母軸插入並固定在扳手主體的孔中。Go to mate tool select rod around face and face of hole relationship should be concentric.

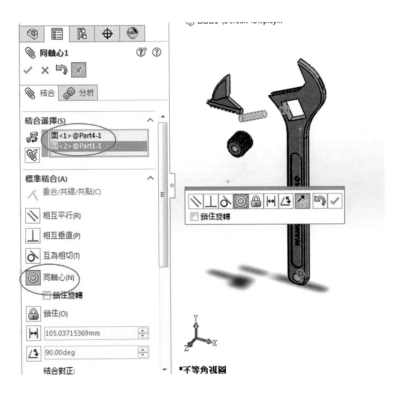

4. 將銷釘面和孔的面設成貼合。Select rod side face and main body's face of hole relationship should be Coincident.

5. 將扳手螺母插入並固定在扳手主體的孔中，請注意不要限制扳手螺母的旋轉。Select the rod around face and inside around face of gear relationship should be concentric.

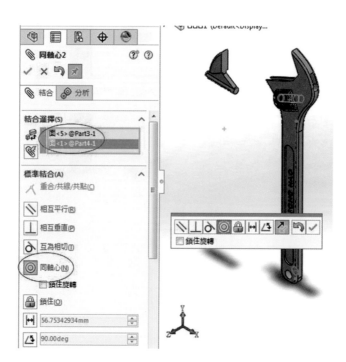

6. 將螺母面和孔的面設成貼合。Select the side face of gear and inside wall face of main body relationship should be Coincident.

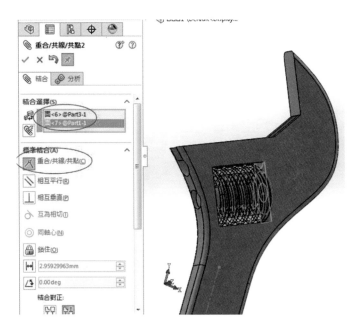

7. 將扳手滑塊的圓柱部分插入扳手主體的相應孔內。Select inside circle face of main body and rod face. Relationship should be concentric.

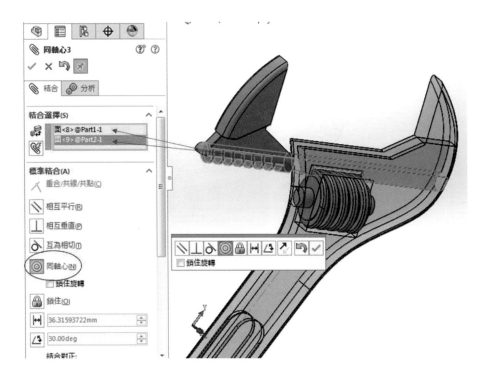

8. 將扳手滑塊的平面部分和扳手主體的相應平面對齊。Select the face of main body and face of part. Relationship should be Coincident.

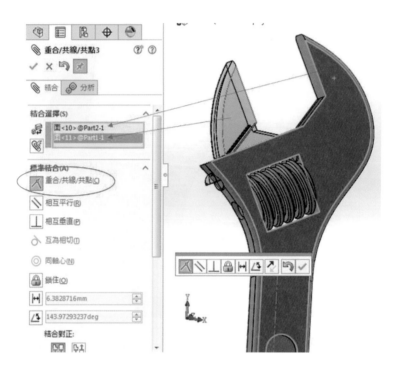

9. 運用相切關係使螺紋和扳手滑塊的面相切。Select the rod teeth face and gear teeth face. Relationship should be tangent.

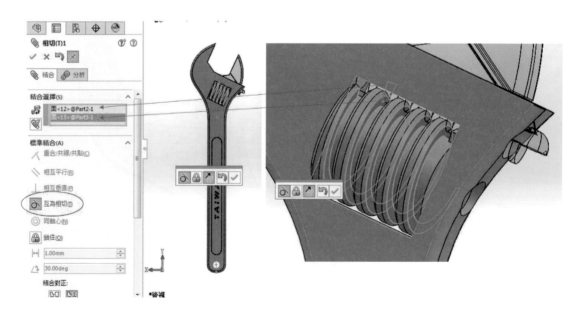

10. 將螺母面和扳手滑塊端點設成貼合。Select gear teeth point and rod teeth face. Relationship should be Coincident.

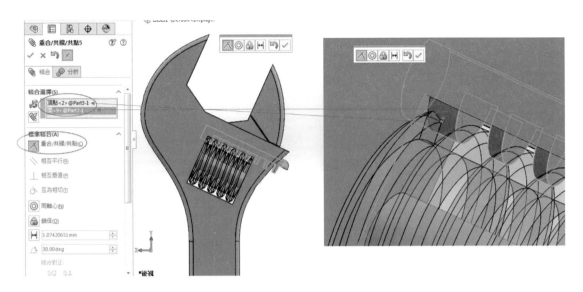

11. 運用測量工具量測扳手可以張開的最大距離，並記錄下來。Use measure tool to measure the maximum length of adjustable spanner. And remember the length.

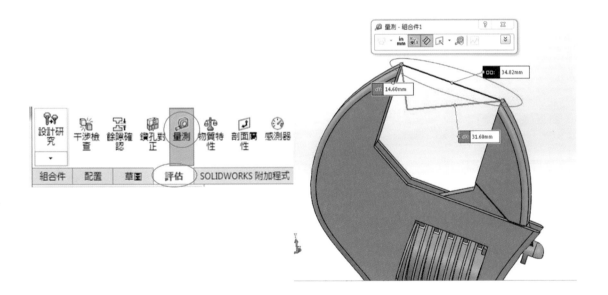

12. 抑制以上兩個結合關係（相切和重合）。Suppress the last 2 steps as shown as below for future steps.

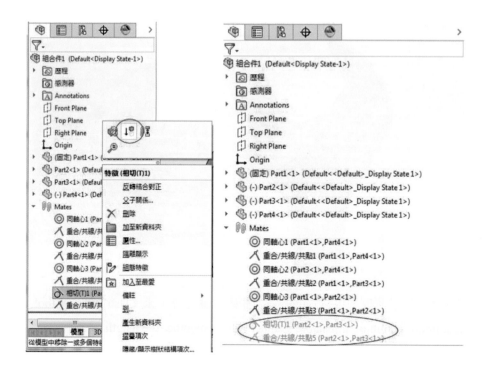

13. 運用進階結合工具，步驟如下圖示。Select the advanced mates and give the length.

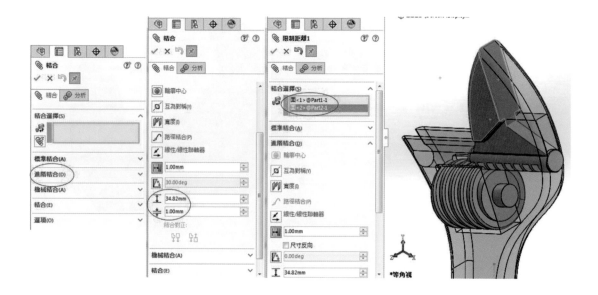

14. 選擇並運用機械結合中的齒輪 - 齒條結合工具，這樣就完成了扳手的組裝。Select mechanical mate and give ratio of movement.

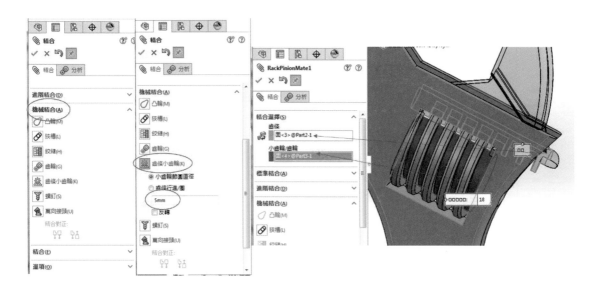

8.14　進階組裝 ──凸輪及彈簧
Advanced Assembly - Cam & Spring

　　在本章中所有的零件已經設計完成，讀者只需從所附光碟片中讀取即可。We only explain the assembly process, the reader can download the attached parts.

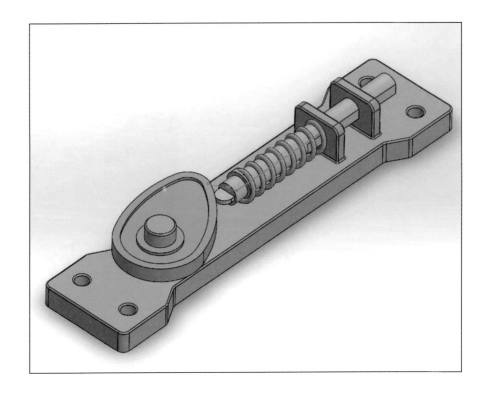

· 組裝

1. 開啟一個新的組合件並插入所有的零件，請注意首先插入基座（BASE），這樣的話基座是固定的。Insert the parts.

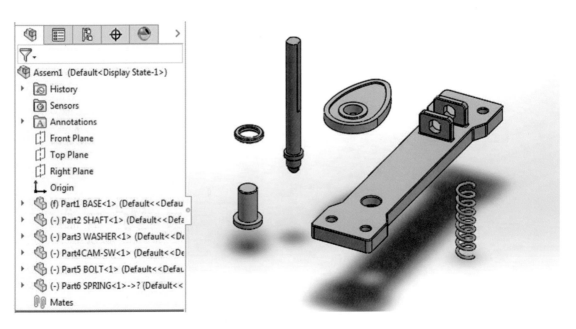

2. 將銷釘插入孔洞如下（同心）。Select the hole of base body (1) and bolt surface (2). Relationship should be concentric.

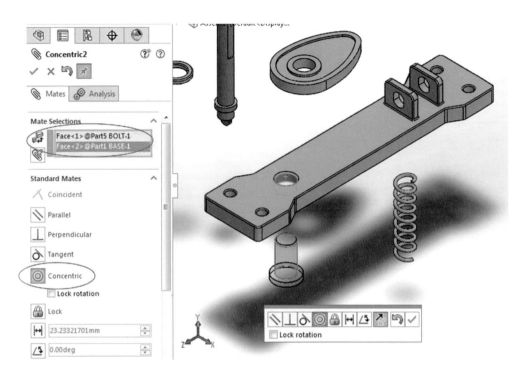

3. 將銷釘插入孔洞如下（貼合）。Select the bolt head back side surface and base body back side surface. Relationship should be coincident.

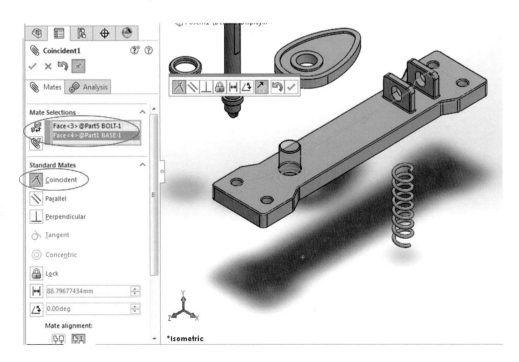

4. 固定墊圈如下（同心）。Select the washer inside surface and bolt surface. Relationship should be concentric.

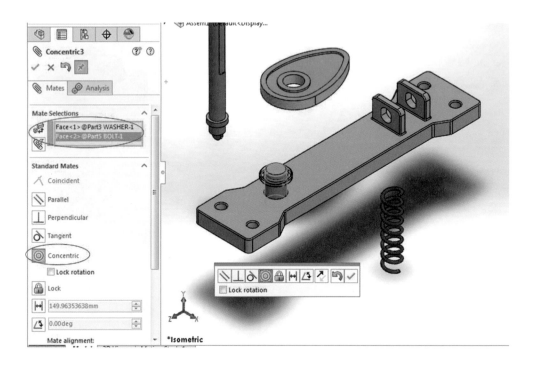

5. 固定墊圈如下（貼合）。Select the washer bottom surface and base body upper surface. Relationship should be coincident.

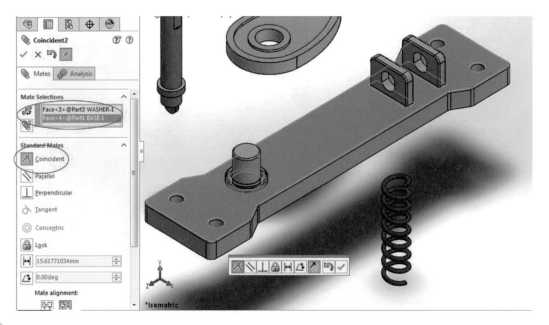

6. 將軸插入孔洞如下（同心）。Select the shaft surface and base body holder inside surface. Relationship should be concentric.

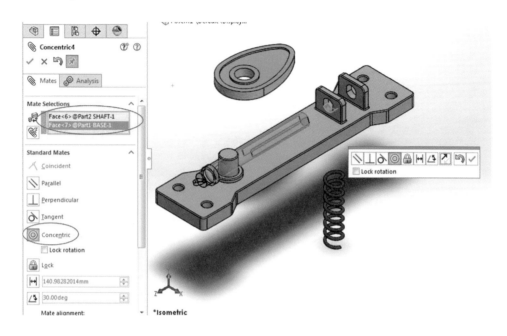

7. 將軸插入孔洞如下（貼合）。Select the shaft flat surface and base body holder side wall. Relationship should be coincident.

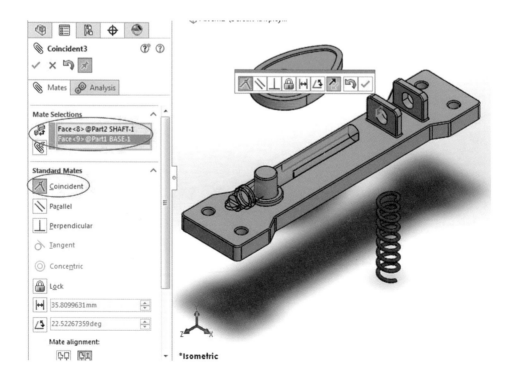

8. 將凸輪裝在銷釘上（同心）。Select the cam hole inside surface and bolt surface. Relationship should be concentric.

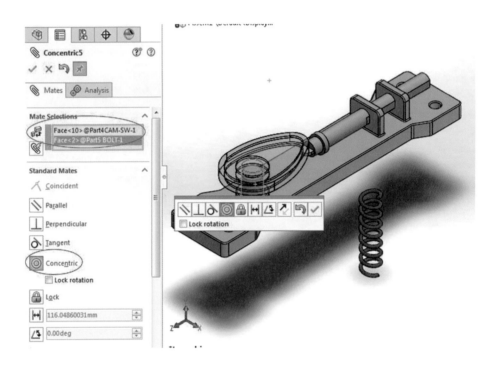

9. 將凸輪裝在銷釘上（貼合）。Select bottom surface of cam's hole and washer upper surface. Relationship should be coincident.

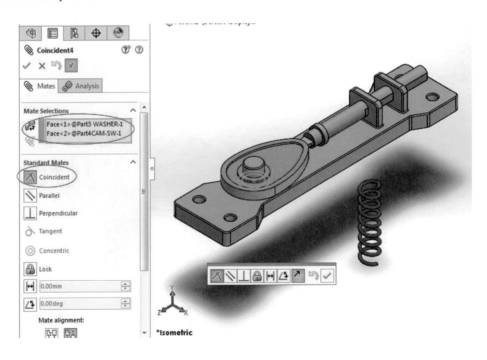

10. 運用機械結合（Mechanical Mates）功能的凸輪（凸輪）選項。Choose the Mechanical Mates. Select the cam's outside surface and shaft end curve as cam follower. After this step shaft should follow the cam movement.

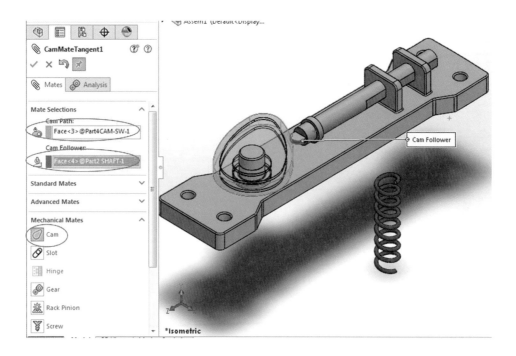

11. 插入彈簧（同心）。Select the vertical line of spring and shaft surface. Relationship should be concentric.

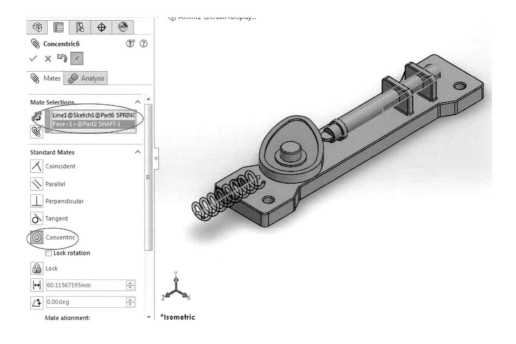

231

12. 點選彈簧中心線並按滑鼠右鍵編輯草圖，這時就會要求輸入關係，選擇彈簧中心線和周的邊線，定義關係為貼合。Select the vertical line of spring and right click then edit sketch. Once click edit sketch it will ask to do relationship. Select the vertical line end point and shaft cylinder edge. Relationship should be coincident.

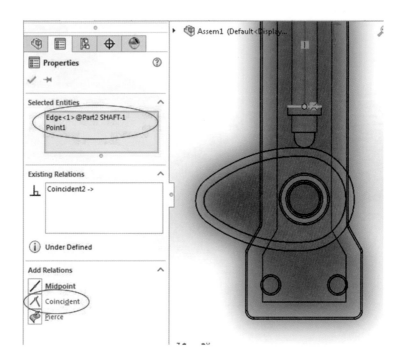

13. 運用智慧尺寸工具定義彈簧中心軸端點和基座上的軸支撐面的間隙為 5mm，然後退出編輯。Use Smart dimension tool to make gap dimension with 5mm. Select vertical line end point of spring and base body holder edge. Then click the exit.

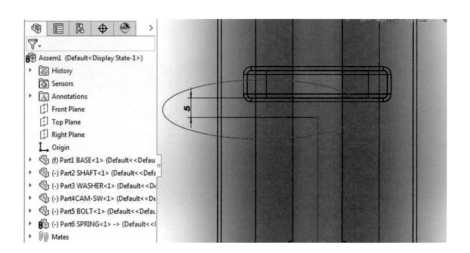

　　這樣我們就完成了所有的組裝步驟，請注意當你轉動凸輪時軸會跟著直線運動，而彈簧不會跟著伸縮，可是在運動模擬時奇跡就會發生。All assembly steps are finished but spring will be not move. After make motion it can move.

・運動模擬 Motion

1. 選擇運動分析，並定義模擬時間為 15 秒。Select the Motion Study tab. Assembly duration 15 seconds.

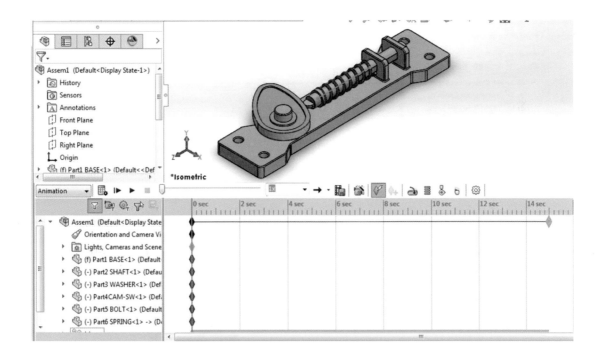

2. 點選動力按鈕，選擇凸輪爲主動件，並定義轉速爲 15RPM。Click motor tool. Select cam cylinder top face and give 15RPM.

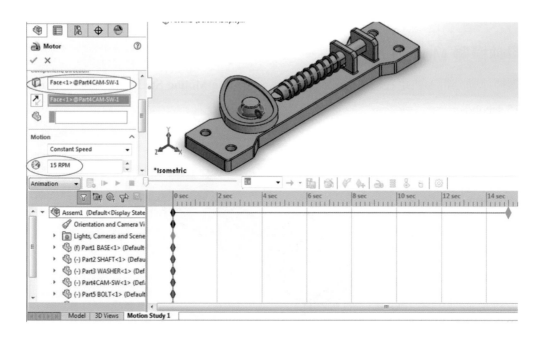

3. 點擊運行按鍵，你就可以觀賞彈簧隨著伸縮的奇跡。滿意的話讀者也可以將影片儲存下來，本書所附的光碟片裡就包含了一個小影片，當然你還可以改變參數並繼續模擬。Play the motion and can save video.

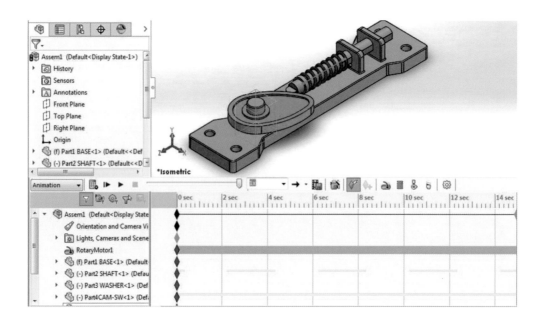

8.15 模具設計基礎 Basics of Mold Design

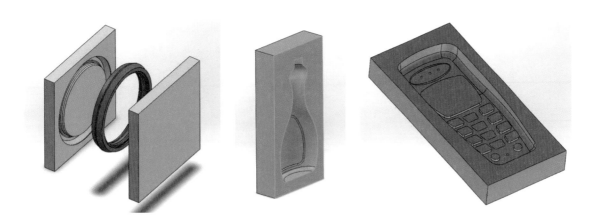

本章節主要介紹利用電腦軟體進行模具設計的基本概念及方法，至於模具滑塊等比較複雜的設計，讀者可以在本章節基礎上參考其他的模具設計書籍及網路教材。The chapter introduce simple Mold design, and basic concepts and Methods.

8.15.1 分模線不明顯的模型 For parts without clear split lines

對於如下分模線不明顯的模型，運用軟體的插入模塑工具便可以非常方便地製作模具。For the model with no clear split lines, an Insert Cavity Feature tool in Solidworks can be applied very easily.

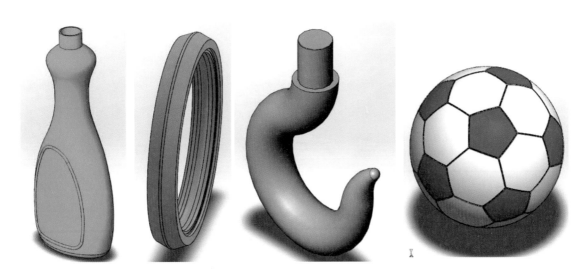

我們以瓶子模具為範例，具體步驟如下：

1. 開啟新的組合件，將模型輸入（瓶子模型在本書所附的 CD 光碟中 8-15-01-Bottle.sldprt），並且找到顯示欲將分模的中心平面（三個步驟如下圖）。

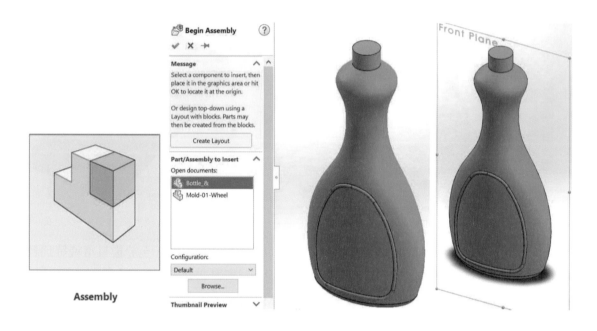

2. 在組合件中插入新零件（Insert/Component/New Part），因為上面已經指向模型的中心平面，現在軟體就主動要求你繪製一個草圖（二個步驟如下圖）。

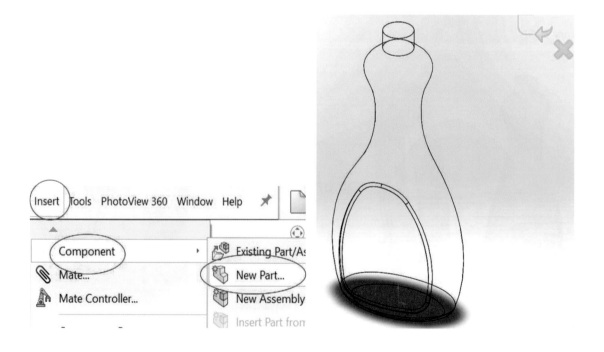

3. 在此中心平面上繪製模仁的截面長方形，以此草圖拉伸成為一半模仁的實體（二個步驟如下圖）。

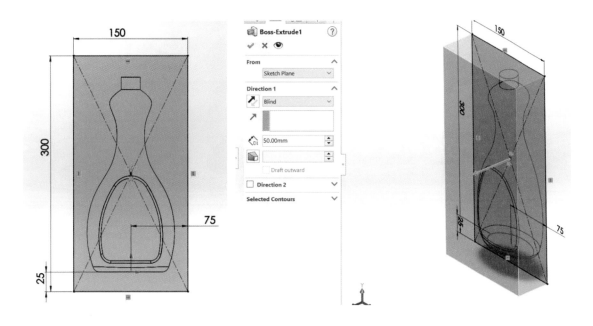

4. 插入 insert－特徵 feature－模塑 cavity，選擇設計模型（二個步驟如下圖）。

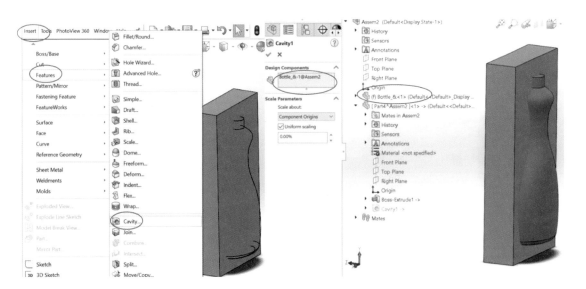

5. 在組合件中將模仁零件打開，就得到了一半的模具（二個步驟如下圖）。Open the assembly you can get the design of the mold.

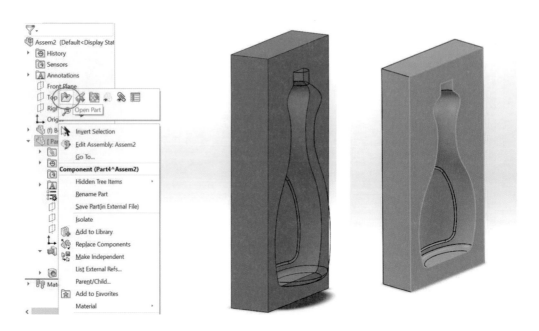

6. 讀者可以用完全相同的方法製作另一半的模具。The reader can use the same way to get the other half of the mold.

　　以下自行車輪胎的模型（模型在本書所附的 CD 光碟中）也可以用完全相同的方法方便地設計出模具。The following mold of a bicycle tire can be made with the same method.

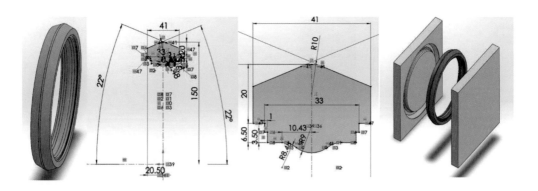

8.15.2 分模線明顯的模型 For parts with clear split lines

　　對於具有明顯分模線的模型，運用軟體提供的模具工具便非常方便。For model with clear split line, the Solidworks mold design tool is recommended.

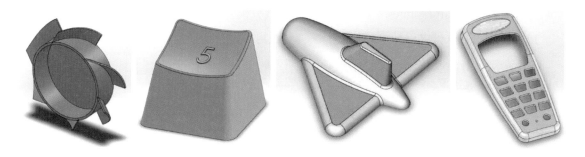

這裡我們以手機機殼為範例，具體步驟如下：Here we use the mold for cellphone cover as an example:

1. 開啟手機機殼模型（模型在本書所附的 CD 光碟中 8-15-02-CellPhone.sldprt）。
2. 開啟模具工具（View/Toolbars/Mold Tools），特徵樹裡展示了已經完成的五個步驟，分別敘述。

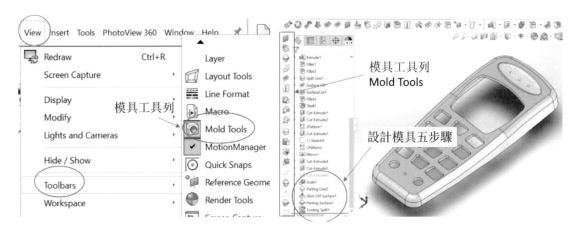

3. 將原始模型進行縮放以考慮材料的縮水率，這裡取 3/1000 也就是將模型放大 1.003 倍，這樣的話當成品冷卻到室溫時可以達到實際設計的尺寸。

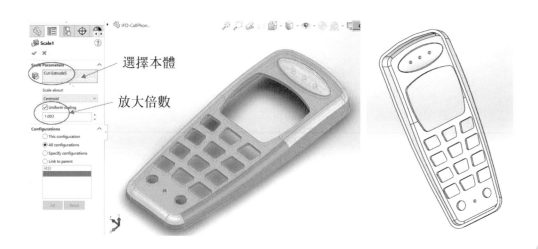

4. 定義分模線（Parting lines）：首先選擇底面為分模方向，再點擊拔模分析按鈕，對於簡單邊界，電腦軟體會自動選擇分模線，讀者也可以自行選擇及編輯。

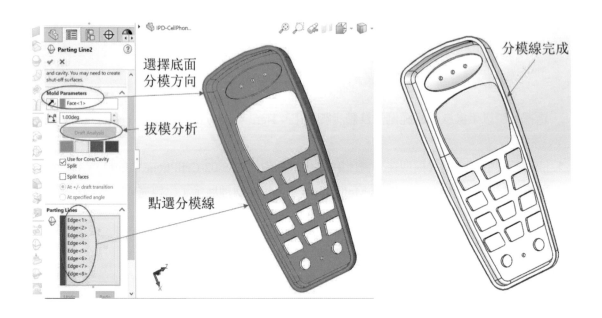

5. 若模型有上下通孔的話就必須建立封閉曲面（Shut-off Surfaces），若沒有的話（以上風扇、鍵盤鍵、模型飛機）就不用做此步驟。對於簡單的孔邊界，電腦軟體會自動選擇，讀者也可以自行選擇及編輯。

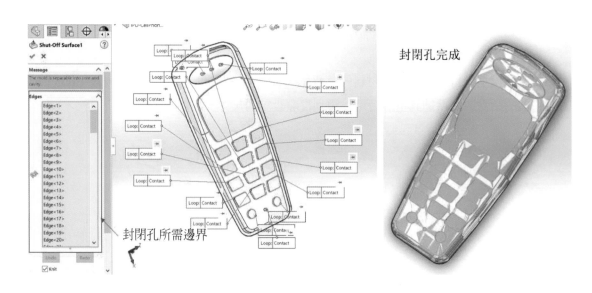

6. 定義分模面（Parting surface）：選擇剛剛定義的分模線（Parting lines），根據模仁大小的需要定義分模面的寬度。

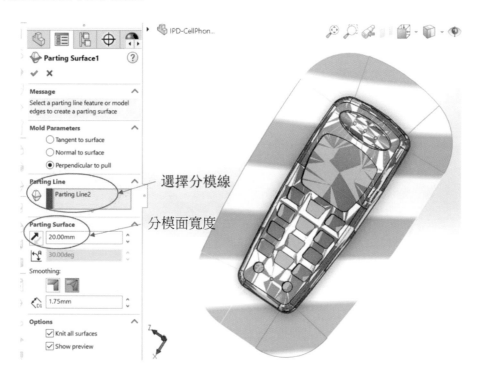

選擇分模線

分模面寬度

7. 模具分割（Tooling split）：在分模面上繪製模仁長寬尺寸，然後再定義上下模仁的高度（右下圖的黃色線段為模仁大小的預覽），這樣就完成了模具的設計。

模具分割工具

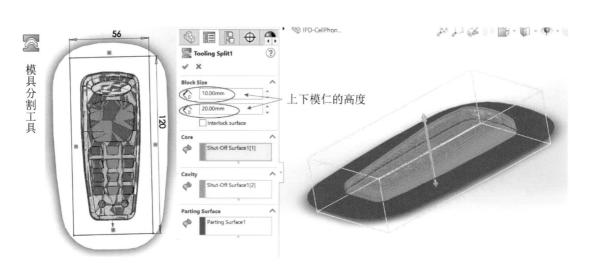

上下模仁的高度

8. 取得上模：在特徵樹上面部分，打開實體（Solid bodies）資料夾，裡面已經存在三個實體模型（原始模型／上模／下模），在上模實體部分滑鼠點擊右鍵，選擇輸入到新的零件（Insert into new part）。

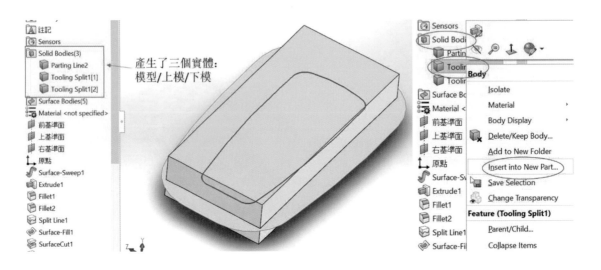

9. 電腦向你確認時，點選 Yes，上模零件就這樣產生了。

10. 讀者可以同樣方法獲得下模零件然後組合確認。

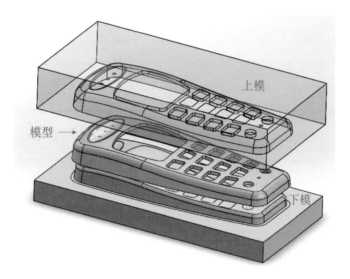

8.16　商標設計 Logo design

8.16.1 商標設計的重要性 The importance of Logo design

　　在現代社會，公司在初建時往往都需要一個商標，以示其公司的特徵。一個好的商標設計需要考慮以下幾點。

1. **引起客人注意 Grab Attention**：Company have few seconds to convince potential customers cause consumers attention spans are too short nowadays. A logo can quickly grab viewers' attention and communicate a company's core values in an interesting way and solid logo can speak on behalf of your company.

2. **第一印象 First Strong Impression**：The logo is first impression and first introduction to consumers. If designed well, it can pique the interest of the public and seek to learn more about the company, in conversely, alienate a potential customer base and basically loss your business. This first impression is your way to immediately express basis for Social participation over the product you sell or niche you dominate the market.

3. **品牌定位基礎 Brand Identity Foundation**：Logo is your company special part of the company brand, which can make winner and loser. Shapes, Art-line, colors, tones and fonts all of this is determined by the story you're trying to tell market or society. Company logo onto all

of your branding materials – letterheads, business cards, landing pages, promotion and can be your last name – creating a concrete, marketable brand identity.

4. 容易記憶 **Memorable**：Logos are a point of identification; they are the symbol that customers use to recognize your brand but the most essential thing is making customer's own strong feelings then people will not forget and already installed in their mind. But people will likely forget your company running a business but they will immediately associate your logo wither their memories of your brand cause this sense is human nature. And the well-designed logo is a visual, aesthetically pleasing element, it triggers positive recall about your brand.

5. 與競爭者區分 **Distinguish from Competition**：Dare to be different with your logo, because your company logo tells consumers why your business is unique. Sure, maybe there are 50 other coffee shops in your city, but *yours* is the only one that's committed to sustainability, and your green, earthy logo drives that message home. Logo express company's value, competency, capability and background to their mission (entertainment, efficiency, and innovation) through the right icon or proper font. Consumer feels - you're better.

6. 品牌忠誠度 **Brand Loyalty**：As your brand grows, your logo is going to become more familiar to a wide range of consumers, and this familiarity creates the perception that you're trustworthy and accessible and make consumers crave consistency. Once they prefer your brand, your customers are going to seek you out again and again – and your logo is the thing they'll look for first.

7. 客戶接受度 **Customer acceptance**：Logo is the first and front part to demonstrate your business and reach consumer feelings. Then create opportunity, bright future through marketing materials such as business cards, flyers, advertisements, etc. If you don't have a logo (and one that stands out), then you are missing an opportunity to make your business stick in the minds of your audience.

8.16.2 電腦輔助商標設計 Computer-aided logo design

市面上有不少軟體可以用來進行商標設計，本書主要介紹 Adobe Illustrator 的使用。

There are several software for logo design on the market, however, Adobe Illustrator is used in this book. Adobe Illustrator is a vector-based design, drawing and graphics editing program published by Adobe in 1987 for the Apple Macintosh and later versions are runs on Microsoft Windows, nowadays using this program all over the world by millions of designers, artists and even ordinary people to create their own gorgeous creations. The program itself is very useful for designing logos, symbols, patterns for fashion, icons, clip-art for artist, blueprints, web, video & film, basic RGB and other precise, resolution-independent illustrations. One of the powerful, superb and efficient usage of illustrator program is design company logos. Therefore if you have capability to create your own logo, this program is definitely designed for you maybe to make you professional illustrator so your future is in your hand.

電腦繪圖基於兩類邏輯：向量及點陣（Vector and Raster）

Computer graphics is based on two appearances, Vector and Raster. Vector are made up of points, lines, and curves that can be infinitely scaled without any loss in image quality. Raster are made up of a set grid of dots called pixels, where each pixel is assigned a color value. Unlike a vector image, raster images are resolution-dependent. Therefore, Vector image will help us more accurately edit, create and import and export creation.

Vector（向量）：Adobe Illustrator creates vector graphics, which are composed of lines and curves defined by mathematical objects called "vectors." Vectors describe a graphic according to its geometric characteristics. For example, a bicycle tire in a vector graphic is drawn using a mathematical equation for a circle with a certain radius, set at a specific location, and filled with a specific color. You can move, resize, or change the color of the tire without losing graphic quality because the underlying equations will compensate for your actions.

A vector graphic is resolution-independent, such as it can be scaled to any size and printed on any output device at any resolution without losing its detail or clarity. As a result, vector graphics are the best choice for type (especially small type) and bold graphics that must retain crisp lines when scaled to various sizes.

本文使用 Adobe illustration 輸出檔案格式為 AI，可是也可以儲存為 eps, pdf, jpeg 等各種圖像格式。

File format is in AI format but it is possible to save files as .eps (encapsulated postscript) and export graphics to.pdf, jpeg, and other graphics formats, although some or all vector editing information may be unavailable if those files are imported back into Illustrator.

程式安裝 **Program installation**：You can find out a program from adobe.com, they provide you 4 types of option Adobe Illustrator Single App Get Illustrator as part of Creative Cloud for just US$20.99/month, Creative Cloud All Apps Get Illustrator and the entire collection of creative apps for US$52.99/month, **Students and teachers** Save over 60% on the entire collection of Creative Cloud apps US$19.99/month. Business Industry-leading creative apps with simple license management starting at US$33.99/month. Also, they have **7 days free trailer** to attempt the illustration program.

商標設計的成本 **Logo design cost：**Whoever is reading this book, here is additional information about logo design cost. A logo design cost can range from $15 to $15,000 and even way more. Because logo design consumes time, patience, innovation, talent, skill and so forth. A logo design can be so expensive because the process of achieving a top-quality logo takes time and talent, and a great logo can help a business succeed. For example, everyone knows that the Pepsi logo reaches $1,000,000. This means the logo can be the property of the company. Therefore, it is suggested that if you have an aim to learn logo design tools such as adobe illustration, you have such a good opportunity to earn an immeasurable amount of money.

8.16.3 開始使用軟體 Getting started in Illustrator

產生新檔 Create New Document. **因為本書所有的商標設計介面都是英語，因此以下步驟均以英文注釋。**

Go File → New Document to create your first document. Type in a Name for the document and click Advanced to select RGB for Color Mode as we will be working for the screen. Click Ok to open a new document.

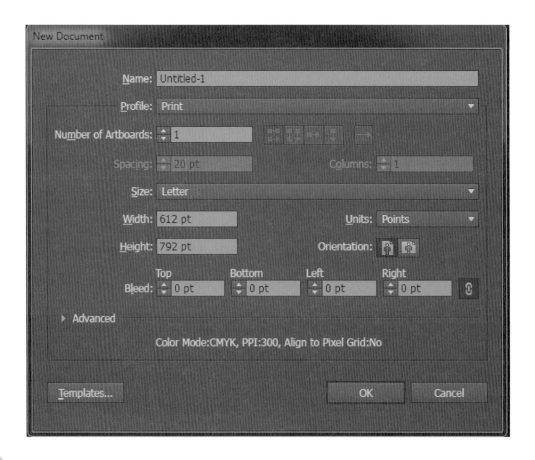

工作畫面 Illustrator Workspace

Below is the workspace and some common terms for calling it. If you are using the latest Adobe Illustrator CC2015, you will notice to a new design for Floating Palette. The rest looks about the same.

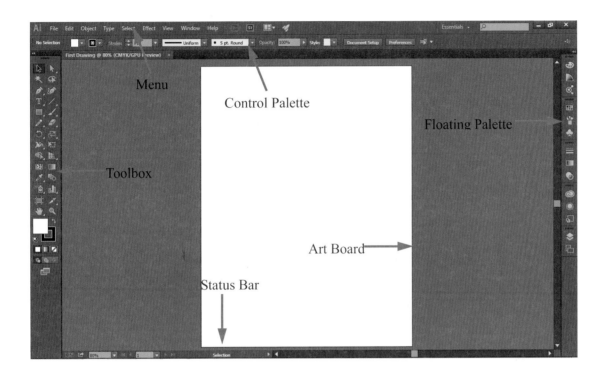

工具箱 Toolbox

This is the handy toolbox which we will use most often. By default, it comes in one single column as shown in the screenshot above. To switch it back to the old 2 columns toolbox, you can simply click the top left mini arrow to toggle it into 2 columns. Some of the tools like Rectangle have more tools hidden. To expand, just click and hold the icon to reveal all the similar tools under that group.

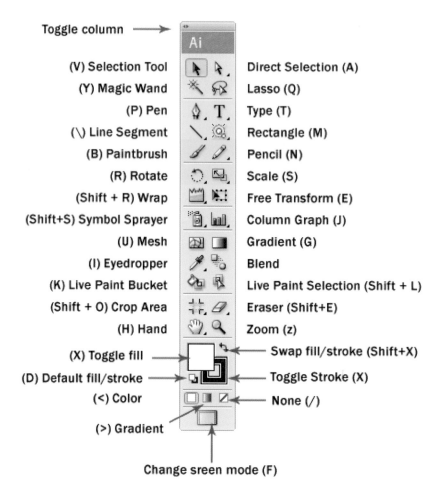

浮動操作板 Floating Palette

This is the floating palette which contains properties for our shapes. It is commonly used for changing colors and stroke width.

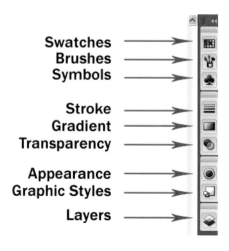

儲存檔案 Saving AI Files

Let's select the type and click on the art-board and type some text. After that go to File>Save. Select Adobe Illustrator (*.AI) for file type and name it "First Drawing". Click Ok after that. Leave the rest of the settings at default. You have successfully saved your first file. To make any more changes you can simply open the Adobe Illustrator file.

8.16.4 實際範例操作 Real Case Practice

接下去用兩個著名商標的圖形範例來進行實際操作，Google Chrome 以及 Food Panda。

8.16.a Google Chrome 商標設計 Google Chrome Logo

每天上班一開機上網首先呈現的就是 Google Chrome 商標，因此本書也來湊個熱鬧，只是試圖運用 Adobe Illustrator 來嘗試繪製此滿街忙碌的可愛商標，沒有絲毫要進行任何商業活動的意思。

由於軟體的英文版本緣故，作者在此就沒有將所有指令翻譯成中文，抱歉。但是讀者應該可以根據以下圖示及步驟容易地學會使用。

打開 Adobe Illustrator 軟體→開啟新檔案→輸入檔案名稱 "Google Chrome Logo " → Run the Adobe Illustrator CC version → Open the New Document → Give the name: "Google Chrome Logo " → Profile → Print, Web, Device, Video& Film and Basic RGB (Choose one of them depending on your purpose)

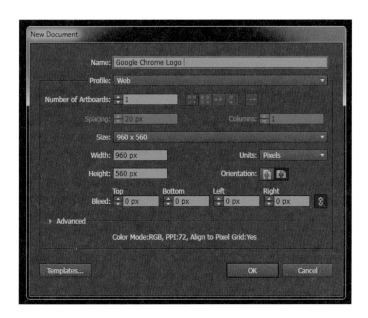

複製商標圖案並貼上在軟體圖面。

Copy the company logo image from the source and paste it inside the software. (https://images. app.goo.gl/fEdgpbfrYVHyWiXr8)

Go to Edit → Paste → Keyboard Shift button and push & pull mouse scroll (If the picture is too small and over big can scale. This function will help to save image proportion)

繪製外圓。

Draw outside circle: Use **Ellipse tool and press shift** from top intersection to bottom edge (Press Shift will give you perfect ellipse and same dimension as logo) Or Double click → set the dimension of Ellipse.

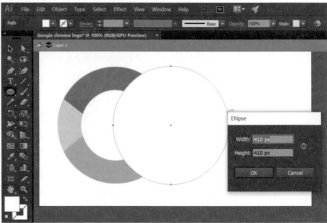

Use **Stroke** and **Fill** function to color up ellipse: Select Ellipse → Double clink on the stroke → Select color

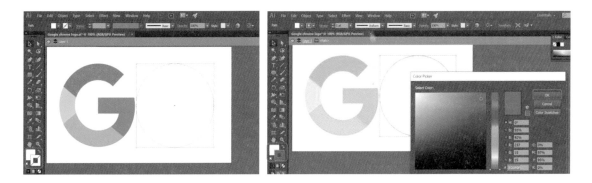

Draw inner circle: **Ellipse tool and press shift** from inside intersection to the bottom edge.

Use **Stroke** and Fill function to color up ellipse: Select Ellipse → Double clink on the stroke → Select color

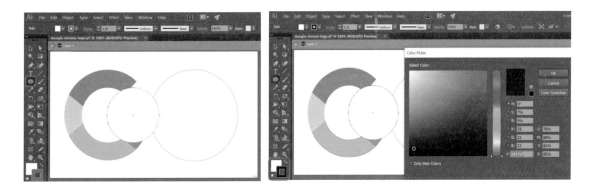

Select two Ellipses → Align → **Vertical & Horizontal** align Center functions. Or match two ellipses center by dragging.

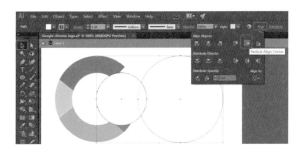

Match two selected Ellipses with original logo by dragging or use align tools → Change **Opacity** 100% - 25% (Which make it transparent)

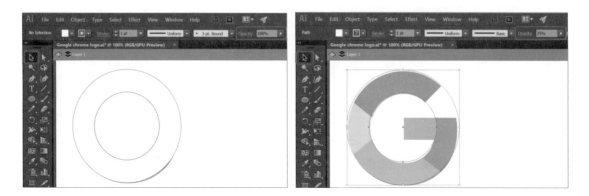

Select the **Rectangle tool** to draw on the logo rectangle shape → **Fill** with color

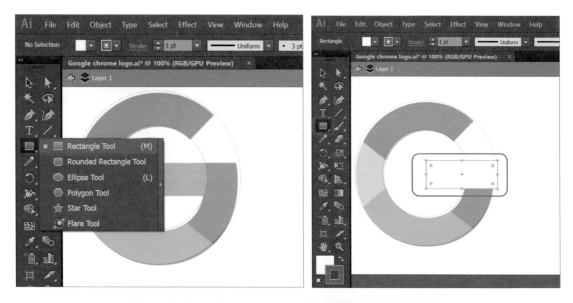

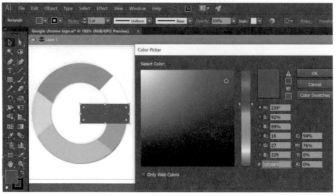

Select the **Pen tool** → draw the split line to separate Ellipse in three shapes.

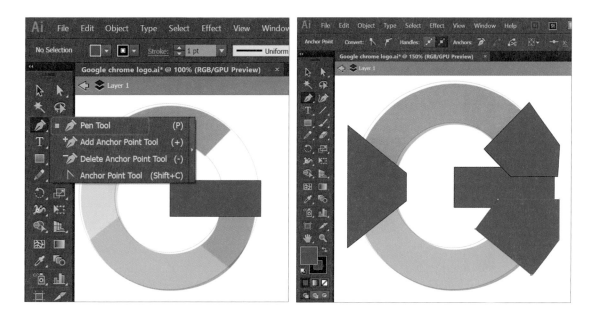

Select the three shapes to make transparent use **Opacity** 100% to 25%. To make sure your separation is correct or not.

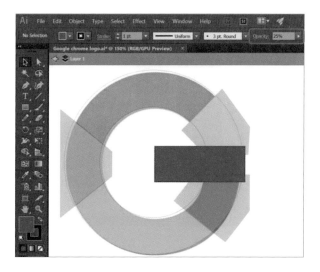

Move the logo image to beside of the drawings → Select Drawing parts **Opacity** change into 100%

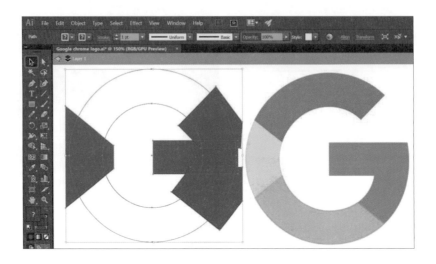

Select all drawing parts → **Pathfinder** → **Divide (if Pathfinder does not exit** → **Shift+Ctrl+F9)**

Mouse right click on the selected parts → **Ungroup it**

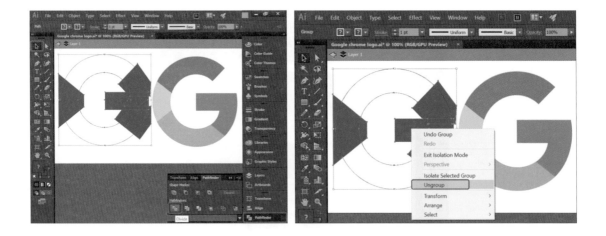

Now can **delete unnecessary parts** → **Select required part** → **Pathfinder** → **Merge**

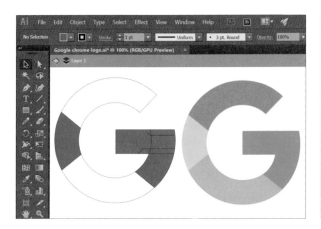 

Use the **Selection tool** to select a part → **Eyedropper tool** → Click on the logo's same parts → pick-up the same color.

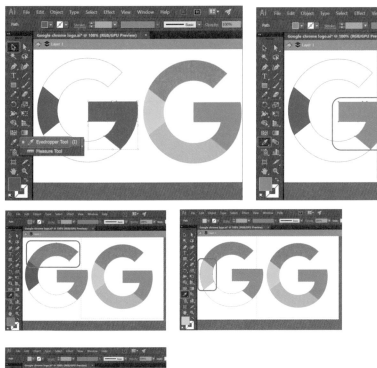

最終我們完成了這個商標的設計。Finally, we have the same logo design.

8.16.b 空腹熊貓商標設計 Foodpanda Logo

空腹熊貓為臺灣人帶來了極大的方便，特別是上班族，因此本書也來湊個熱鬧，只是試圖運用 Adobe Illustrator 來嘗試繪製此滿街忙碌的可愛商標，沒有絲毫要進行任何商業活動的意思。

由於軟體的英文版本緣故，作者在此就沒有將所有指令翻譯成中文，抱歉。但是讀者應該可以根據以下圖示及步驟按部就班地學會使用。

Run the Adobe Illustrator CC version → Open the New Document → Give the name: Foodpanda → Profile → Print, Web, Device, Video & Film and Basic RGB (Choose one of them depending on your purpose)

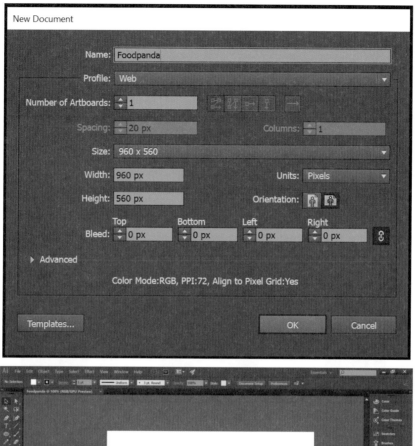

Copy the company logo image from any source (Foodpanda Logo Link: https://images.app. goo.gl/Y6u9iFiabqazop8g6) and paste it inside the software. Go to Edit → Paste or Ctrl +V → Click the keyboard Shift button and push & pull mouse scroll (If the picture is too small and over big can scale. This function will help to keep image proportion)

The Screen right top section shows Foodpanda original logo picture position and Screen left bottom section shows the zoom percent

Draw base rectangle: Select **Rectangle Tool** → draw from the sharp top edge to Bottom same as logo design → **Selection tool** mouse right-click → **Arrange** → **Send to Back**

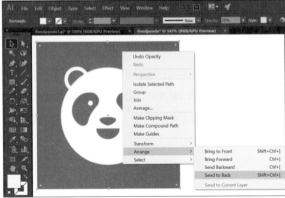

Draw Foodpanda head, ears and eye circles: Select **Ellipse tool** click → set the dimension of Ellipses.

Head Ellipse - 260 px	Ear outside Ellipse – 70 px
Ear inside Ellipse – 50 px	Eye Ellipse – 15 px

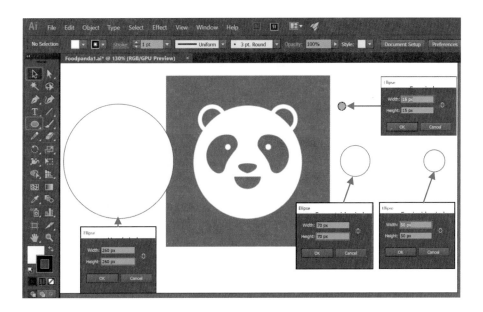

Select four Ellipses transparent: Keyboard **Shift button** select all Ellipses → **Opacity** 100% - 25%

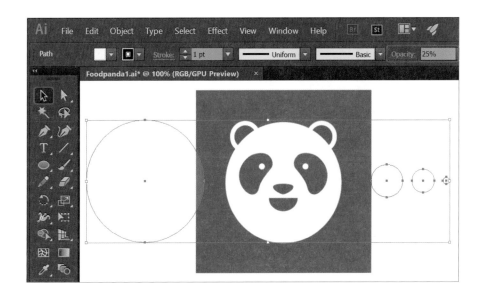

Match the Ellipses with original logo shape by dragging one by one or using align tools.

Copy Ears & Eye Ellipses → Keyboard shortcut **Ctrl+C** → **Ctrl+V**

Select the **Pen tool** → draw the eye & nose shape by manually

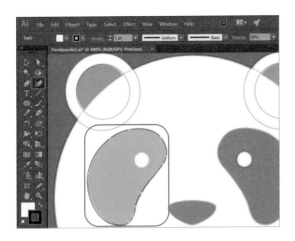 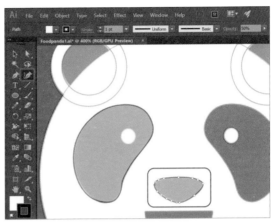

Select the **Pen tool** → draw mouth open curve with 3 points → **Selection Tool** → **Pathfinder** → **Shape Modes** → **Minus front** (Closing mouth open curve – Compound shape)

Duplicate eye shape → **Selection Tool** select the shape → mouse right-click → **Transform** → **Reflect** → select **Vertical** (to flip the eye shape) → **Tick Preview** (review before copy the shape) → **Copy**

 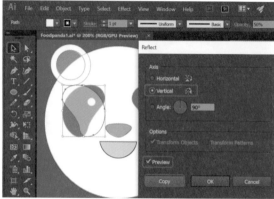

Eye shape overlapped with eye Ellipses → **mouse right-click** → **Arrange** → **Send Backward** (To bring eye ellipses in front)

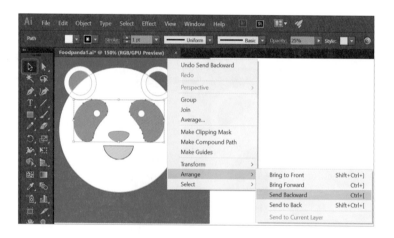

Combine the shape → Select **Ear Ellipses with Head Ellipse** → **Select Shape Builder Tool** → **Drag your mouse** according to the following steps

After use of Shape builder tool **move the original logo image** to beside of the drawings → **can check the drawings completion**, select all drawing parts **Selection Tool** → **Opacity 25% - 100%**

Use **Selection Tool** to select the parts → **Eyedropper tool** → Click on the logo's same parts → pick-up same color.

Remove the **Stroke** Black color which highlighted the parts before.

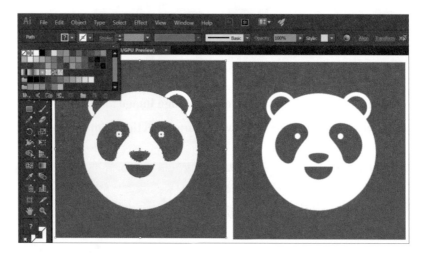

Finally, we have the same logo design and your new creative logo design.

需要的話，讀者可以選擇不同的背景色。

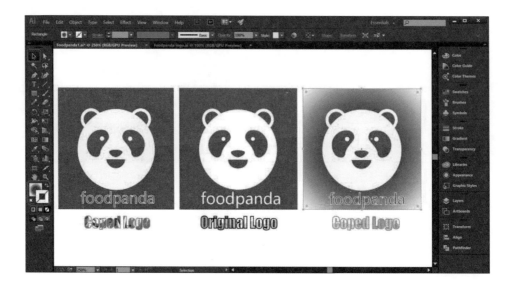

附：更複雜的藝術作品 More complex artworks

實際上，此電腦程式可以進行更複雜的商品設計及藝術創作，以下即是以本書作者的相片爲基礎所創作的不同形態的卡通設計。In fact the software is very flexible and useful. For example the following are the graphs made from AI according to Authors' photographs:

王松浩　　　　　　　　尤瑞雅　　　　　　　　貝德杰

參考資料 References

1. 曾坤明，《工業設計的基礎》，1979 年。

2. Harold Van Doren 著，黃鼎貴、曲新生譯，《工業設計 Industrial Design》，1976 年。

3. 林振陽，《造型（二）》，三民書局，1993 年。

4. 李娟，《商品的誕生》，中國建工。

5. Alvin R. Trlley，《人體工學圖解》。

6. 郭光，《工業產品造型設計》，中國建工。

7. 殷紫，《北歐新銳設計》，重慶出版社。

8. Andrew Davey 著，魯曉波譯，《精細設計》，清華大學出版社。

9. 劉春榮，《人機工程學應用》，上海人民美術出版社。

10. 張淩浩，《產品的語言》，中國建工。

11. 克里斯，萊夫特瑞，《歐美工業設計五大材料》，上海人民美術出版社。

12. 王行凱，《設計廣場系列基礎教材：3 維設計、構成》，工業設計共 3 冊，廣西美術出版社。

13. 王行凱、程惠琴，《工業設計》，廣西美術出版社。

14. 官政能，《產品物徑：設計創意之生成、發展與應用》，藝術家出版社。

15. McArdie, Ann, *Decorating with color and texture*, Rockport Publishers, 2000。

16. Gill, Martha, *Color Harmony*, Rockport Publishers, 2001。

17. 韓春明，《工業產品造型設計》，機械工業出版社。

18. 蘭德，《基本設計》，正元圖書公司。

19. Donald A. Norman 著，桌燿宗譯，《設計心理學 The Psychology of Everyday Things》。

20. BernHard E. Burdek 著，胡佑宗譯，《工業設計──產品造型的歷史、理論及實務》，亞太圖書出版社。

21. 蓋爾・格利特・漢娜著，李樂山、韓琦、陳仲華譯，《設計元素》，中國水利出版社。

22. 王松浩、陳維仁、劉風源，《電腦輔助設計與工具機實列》，五南圖書出版。

23. 王松浩、莊昌霖、熊效儀，《逆向工程技術及實作》，五南圖書出版。

24. http://www.wikipedia.org/

25. https://www.youtube.com/watch?v=4jrSYMrmbyQ

附錄 Appendix

附錄 I. 基本概念思考題 Appendix I: Quiz for Basic Concepts

1. 請解釋 OXM，包括 OEM、ODM、OBM。

   Explain OXMs (OEM, ODM and OBM).

2. 設計的定義。

   Definition of Design.

3. 設計的主要種類。

   Variety of designs.

4. 何謂工業產品？

   Definition of Industrial Products.

5. 工業產品分類。

   Category of Industrial products.

6. 工業產品設計的定義。

   Definition of Industrial product design.

7. 工業產品設計的三要素，請分別詳細說明。

   Three basic components of Industrial Product (details).

8. 產品開發的三個主要部門。

   Three Departments for Product Development.

9. 產品成功的五個關鍵。

   Five Keys for Product success.

10. 典型的新產品設計過程包含四個階段。

    What are the four steps involved in a typical product design process?

11. 對於新產品開發概念會發表其不容忽視的觀點的關鍵團體。

    What are the groups whose opinions are key important to the new product development?

12. 何謂透視圖法、遠近畫法？

    What is Perspective drawing/sketch?

13. 描述基本色彩圓盤，怎樣延展成為色彩球？

    Please sketch the basic color wheel; how to make a color sphere?

14. 簡介色彩數位化原理。

    Explain the principle of color digitization.

15. 請列出五項形態美的原則。

List at least five principles of form of beauty.

16. 黃金比例的幾何以及數學背景，舉出四項實際範例。

Explain the basics of the Golden section (including the numerical series) and list four examples.

17. 描述四項基本矩形（Orthogone）的幾何以及數學形式。

Sketch and write mathematical expression for the four basic Orthogones.

18. 列印紙 A0、A1……的長寬比是？是什麼原理？

What is the ratio for printing papers A0, A1 … and why?

19. 何謂流線型（Stream-Line）？舉出四個實例。

Explain what is Stream-Line and list four examples.

20. 何謂仿生學（Bionics）？舉出四個實例。

Explain Bionics and list four examples.

21. 請解釋人們記憶的基本結構。

Explain the Structure of memory.

22. 例舉出讓你想改進的最重要產品目標（任何產品）。

List the most important product improvement you like to make (any product).

23. 找出你身邊產品在設計上的錯誤或缺陷（任何產品）。

Find at least one product design mistake around you.

附錄 II.　產品課程設計 Appendix II: Single Student Course Project

一、可能的選擇（不局限）Possible choices but not limit

1. 香水瓶／飲料瓶／酒瓶 Bottles
2. 杯子／壺 Cups and Pots
3. 眼鏡 Eye Glasses
4. 滑鼠／鍵盤／電腦（外形）Computer Mouse/Keyboard/Cases
5. 手機／藍芽／電話 Cell Phone/Blue Tooth/Telephone
6. 安全帽／手套／傘 Helmet/Gloves/Umbrella
7. 桌椅 Chair and Table
8. 沙發／床 Sofa/Bed
9. 燈具 Lights
10. 玩具 Toys
11. 汽車（外形）Car
12. 遊艇（外形）Boat
13. 飛機（外形）Plane

二、設計構思和報告主要架構 Design report contents

1. 產品名稱 Title of the Product
2. 主要市場 Major Market
3. 客戶群 Customer Group
4. 產品主要優點／賣點 Advantages and Selling Points
5. 產品主要功能 Major Functions
6. 人機互動考量 Human Interaction Consideration
7. 主要材料 Major Materials
8. 產品草圖 Major Sketches in 3D

國家圖書館出版品預行編目資料

電腦輔助工業產品設計及實作／王松浩、尤瑞
雅、貝德杰著. -- 初版. -- 臺北市：五南
圖書出版股份有限公司, 2021.08
　面；　公分
ISBN 978-986-522-328-1 (平裝)

1.工業設計　2.工業產品　3.電腦輔助設計

440.8029　　　　　　　　　109016490

5F68

電腦輔助工業產品設計及實作

作　　者 ― 王松浩、尤瑞雅、貝德杰

發 行 人 ― 楊榮川

總 經 理 ― 楊士清

總 編 輯 ― 楊秀麗

副總編輯 ― 王正華

責任編輯 ― 張維文

封面設計 ― 姚孝慈

出 版 者 ― 五南圖書出版股份有限公司

地　　址：106台北市大安區和平東路二段339號4樓

電　　話：(02)2705-5066　　傳　　真：(02)2706-6100

網　　址：https://www.wunan.com.tw

電子郵件：wunan@wunan.com.tw

劃撥帳號：01068953

戶　　名：五南圖書出版股份有限公司

法律顧問　林勝安律師事務所　林勝安律師

出版日期　2021年8月初版一刷

定　　價　新臺幣450元